T0249895

Mark S. Ashton
Jennifer L. O'Hara
Robert D. Hauff
Editors

Protecting Watershed Areas: Case of the Panama Canal

Protecting Watershed Areas: Case of the Panama Canal has been co-published simultaneously as *Journal of Sustainable Forestry*, Volume 8, Numbers 3/4 1999.

Pre-publication REVIEWS, COMMENTARIES, EVALUATIONS . . .

"This book, a compilation of papers by graduate students at Yale University gives a different perspective on resource management in the tropics than one may typically expect. Each chapter is based on a week of field observations in Panama, discussing . . . the many on-site resource managers and study of literature. In Panama the main focus is on how to integrate and manage the multiple uses of the watershed that provides water for floating ships through the locks, requiring great quantities daily, and domestic and industrial water. . . . Several Panamanian agencies have the task of developing and implementing a coordinated and effective plan that addresses politi-cal and social issues as well. Each chapter poses an idea for addressing a segment of the task, not necessarily the answer, but a starting point for further inquiry. The literature citings of each chapter are an excellent gateway to information available to stimulate inquiry toward development of an adaptive management plan that will sustain watershed, the people who live there, and the commerce of the world. For those interested in studying and facilitating integrated resource management addressing the ecological, social and political concerns in virtually any watershed this book should be a useful starting point for ideas and references."

Peter R. Hannah, PhD
Professor, Forestry
University of Vermont
School of Natural Resources
Burlington, VT 05405

"**T**he principal objective of this volume is to focus attention on the Panama Canal Watershed Area as a case study in sustainable development and conservation. This Watershed Area is affected by many of the interacting social and ecological forces that shape environmental problems throughout the world. The importance of these problems in this particular case is heightened by (1) its strategic location–much of the water from this region is used to facilitate passage of ships through the Canal locks; (2) the critical time in its history–the management of the Canal and adjoining land is in a period of transition as control is transferred from the United States to the Republic of Panama at the end of 1999. These two factors make the study of far greater importance than if it dealt with an area chosen simply to represent "average" tropical watershed areas.

Many of the chapters intersect several fields of study, including ecology, forestry, sociology, land-use planning, hydrology, economics, and natural resource policy. Some of the most useful chapters are those dealing with agroforestry, carbon sequestration by forests, sedimentation and land use, and ecotourism. These topics have frequently been dealt with elsewhere in broad generalities, in which rather idealized views of management possibilities are commonly presented. In contrast, these chapters address the problems and potential solutions realistically in an analytical fashion, incorporating the details of the Panama situation; this is a refreshing and useful approach. Other chapters deal with broader issues of organizational structure and behavior in the Panamanian and international resource agencies involved in setting policy in the Watershed Area; these are of great concern, as U.S. influence is removed from the decision-making process.

This book also serves an additional important function as a guide for developing graduate courses in integrated natural resource management. The chapters were written by graduate students in the Yale School of Forestry and Environmental Studies who were participating in a semester-long course which included a 10-day field trip to Panama. The chapters are professional in quality . . . the treatment of each topic is thorough and insightful."

Matthew J. Kelty, PhD
Associate Professor
Department of Forestry
and Wildlife Management
University of Massachusetts

More pre-publication
REVIEWS, COMMENTARIES, EVALUATIONS . . .

"**A**ctually, the text has much to recommend it and, . . . it makes a valuable contribution to the literature on conservation and development in the neo-tropics.

. . . Using a combination of bio-physical and ecological assessments to refocus attention to the Canal *watershed*, these student writings provide a fresh yet realistic account of the landscape that Panama will receive, together with a set of policy recommendations for resource management agencies if the former Canal Zone is not to become a millstone around Panama's neck. . . . this particular collection of reports focused on the future of the Panama Canal Watershed . . . Viewed through the eyes of the next generation of professional resource managers, the Canal Watershed represents a unique opportunity to test and apply models of conservation and development on a landscape where American institutions simply cannot afford to fail. The interests of both North and South converge in a serendipitous way on this tiny slice of land that we believe is vital to development interests and hope is vital to conservation interests. Can we learn from the Yale students' writings? Absolutely, but not because they are necessarily correct in all their assessments, but because they force us to reconsider Panama (and the world) as it *should* be, not just for elites but for all Americans, North and South."

Raymond P. Guries
Professor of Forestry
Department of Forestry
University of Wisconsin at Madison
Madison, Wisconsin

Protecting
Watershed Areas:
Case of the Panama Canal

Protecting Watershed Areas: Case of the Panama Canal has been co-published simultaneously as *Journal of Sustainable Forestry,* Volume 8, Numbers 3/4 1999.

The *Journal of Sustainable Forestry* Monographic "Separates"

Below is a list of "separates," which in serials librarianship means a special issue simultaneously published as a special journal issue or double-issue *and* as a "separate" hardbound monograph. (This is a format which we also call a "DocuSerial.")

"Separates" are published because specialized libraries or professionals may wish to purchase a specific thematic issue by itself in a format which can be separately cataloged and shelved, as opposed to purchasing the journal on an on-going basis. Faculty members may also more easily consider a "separate" for classroom adoption.

"Separates" are carefully classified separately with the major book jobbers so that the journal tie-in can be noted on new book order slips to avoid duplicate purchasing.

You may wish to visit Haworth's website at . . .

http://www.haworthpressinc.com

. . . to search our online catalog for complete tables of contents of these separates and related publications.

You may also call 1-800-HAWORTH (outside US/Canada: 607-722-5857), or Fax 1-800-895-0582 (outside US/Canada: 607-771-0012), or e-mail at:

getinfo@haworthpressinc.com

Protecting Watershed Areas: Case of the Panama Canal, edited by Mark S. Ashton, Jennifer L. O'Hara, and Robert D. Hauff (Vol. 8, No. 3/4, 1999). *"This book makes a valuable contribution to the literature on conservation and development in the neo-tropics. . . . These writings provide a fresh yet realistic account of the Panama landscape." (Raymond P. Guries, Professor of Forestry, Department of Forestry, University of Wisconsin at Madison, Madison, Wisconsin)*

Sustainable Forests: Global Challenges and Local Solutions, edited by O. Thomas Bouman and David G. Brand (Vol. 4, No. 3/4 & Vol. 5, No. 1/2, 1997). *"Presents visions and hopes and the challenges and frustrations in utilizations of our forests to meet the economical and social needs of communities, without irreversibly damaging the renewal capacities of the world's forests." (Dvoralai Wulfsohn, PhD, PEng, Associate Professor, Department of Agricultural and Bioresource Engineering, University of Saskatchewan)*

Assessing Forest Ecosystem Health in the Inland West, edited by R. Neil Sampson and David L. Adams (Vol. 2, No. 1/2/3/4, 1994). *"A compendium of research findings on a variety of forest issues. Useful for both scientists and policymakers since it represents the combined knowledge of both." (Abstracts of Public Administration, Development, and Environment)*

Protecting
Watershed Areas:
Case of the Panama Canal

Mark S. Ashton
Jennifer L. O'Hara
Robert D. Hauff
Editors

Protecting Watershed Areas: Case of the Panama Canal has been co-published simultaneously as *Journal of Sustainable Forestry*, Volume 8, Numbers 3/4 1999.

Food Products Press
An Imprint of
The Haworth Press, Inc.
New York • London • Oxford

Published by

Food Products Press®, 10 Alice Street, Binghamton, NY 13904-1580

Food Products Press® is an imprint of The Haworth Press, Inc., 10 Alice Street, Binghamton, NY 13904-1580 USA.

Protecting Watershed Areas: Case of the Panama Canal has been co-published simultaneously as *Journal of Sustainable Forestry*, Volume 8, Numbers 3/4 1999.

The development, preparation, and publication of this work has been undertaken with great care. However, the publisher, employees, editors, and agents of The Haworth Press and all imprints of The Haworth Press, Inc., including The Haworth Medical Press® and Pharmaceutical Products Press®, are not responsible for any errors contained herein or for consequences that may ensue from use of materials or information contained in this work. Opinions expressed by the author(s) are not necessarily those of The Haworth Press, Inc.

Cover design by Thomas J. Mayshock Jr.

Library of Congress Cataloging-in-Publication Data

Protecting watershed areas: case of the Panama Canal / Mark S. Ashton, Jennifer L. O'Hara, Robert D. Hauff, editors.
      p. cm.
     "Co-published simultaneously as Journal of sustainable forestry, volume 8, numbers 3/4 1999."
    Includes bibliographical references.
    ISBN 1-56022-064-3 (alk. paper)–ISBN 1-56022-066-X (alk. paper)
    1. Ecosystem management–Panama–Panama Canal Watershed. 2. Natural resources–Panama–Panama Canal Watershed–Management. 3. Land use–Panama–Panama Canal Watershed. I. Ashton, Mark S. II. O'Hara, Jennifer L. III. Hauff, Robert D.
QH77.P357 P76 1999
333.73'097287–dc21
                                                 99-048247

INDEXING & ABSTRACTING

Contributions to this publication are selectively indexed or abstracted in print, electronic, online, or CD-ROM version(s) of the reference tools and information services listed below. This list is current as of the copyright date of this publication. See the end of this section for additional notes.

- *Abstract Bulletin of the Institute of Paper Science and Technology*
- *Abstracts in Anthropology*
- *Abstracts on Rural Development in the Tropics (RURAL)*
- *AGRICOLA Database*
- *Biostatistica*
- *BUBL Information Service, an Internet-based Information Service for the UK higher education community*
- *CNPIEC Reference Guide: Chinese National Directory of Foreign Periodicals*
- *Engineering Information (PAGE ONE)*
- *Environment Abstracts*
- *Environmental Periodicals Bibliography (EPB)*
- *Forestry Abstracts; Forest Products Abstracts (CAB Abstracts)*
- *GEO Abstracts (GEO Abstracts/GEOBASE)*
- *Human Resources Abstracts (HRA)*
- *Journal of Planning Literature/Incorporating the CPL Bibliographies*
- *Referativnyi Zhurnal (Abstracts Journal of the All-Russian Institute of Scientific and Technical Information)*
- *Sage Public Administration Abstracts (SPAA)*
- *Sage Urban Studies Abstracts (SUSA)*
- *Wildlife Review*

(continued)

Special Bibliographic Notes related to special journal issues
(separates) and indexing/abstracting:

- indexing/abstracting services in this list will also cover material in any "separate" that is co-published simultaneously with Haworth's special thematic journal issue or DocuSerial. Indexing/abstracting usually covers material at the article/chapter level.

- monographic co-editions are intended for either non-subscribers or libraries which intend to purchase a second copy for their circulating collections.

- monographic co-editions are reported to all jobbers/wholesalers/approval plans. The source journal is listed as the "series" to assist the prevention of duplicate purchasing in the same manner utilized for books-in-series.

- to facilitate user/access services all indexing/abstracting services are encouraged to utilize the co-indexing entry note indicated at the bottom of the first page of each article/chapter/contribution.

- this is intended to assist a library user of any reference tool (whether print, electronic, online, or CD-ROM) to locate the monographic version if the library has purchased this version but not a subscription to the source journal.

- individual articles/chapters in any Haworth publication are also available through the Haworth Document Delivery Service (HDDS).

Protecting Watershed Areas: Case of the Panama Canal

CONTENTS

ABOUT THE EDITORS

Mark S. Ashton, PhD, ME, is Associate Professor of Silviculture, and Director of the Tropical Resources Institute at Yale University School of Forestry and Environmental Studies in New Haven, Connecticut. Professor Ashton conducts research on the biological and physical processes governing the regeneration of national forests and on the creation of their agroforestry analogs. He is the author of three books concerning silviculture practices of forests in temperate and tropical realms and numerous journal articles. The results of his research have been widely applied to the development and testing of techniques for restoration of degraded lands and for the management of forest lands for human inhabitants.

Jennifer L. O'Hara, MFS, BA, is a doctoral candidate at Yale University in the School of Forestry and Environmental Studies, New Haven, Connecticut. Her dissertation focuses on the development of a monitoring approach for the harvest of nontimber forest products. Ms. O'Hara taught a graduate seminar in protected areas management at Yale University and over the last seven years she has conducted research and assisted local nonprofit initiatives in Belize. Her fields of interest include protected areas management and ecosystem ecology.

Robert D. Hauff, MF, BA, is a research ecologist in Micronesia for the USDA Forest Service Institute of Pacific Islands Forestry where he conducts interdisciplinary research on mangrove swamps. In 1998, Mr. Hauff carried out summer research in the Peruvian Amazon funded by the Tropical Resources Institute.

Foreword

In a number of publications and different forums, watersheds are presented as adequate sustainable development units. However, watersheds vary in size, and in complexity, complexity defined in terms of different land use and functions expected from it. The Panama Canal Watershed covers approximately 3% of Panama's territory, and is perhaps our most complicated challenge in watershed management. Within its 365,000 hectares, different activities take place: agriculture, cattle raising, tourism, conservation of five national parks, reforestation, fishing, port services, suburban developments hydro-electric plants, potable water plants, and land and aquatic inter-oceanic transportation, to name a few. Needless to say, the whole of these activities represents much of the economic activity of the country. Its inhabitants include a mosaic of indigenous and campesino communities, as well as the two biggest cities of the Republic, Panama City and Colon, which hold approximately half the population of Panama.

It was not until the 1980's that the watershed concept was introduced to the management of natural resources in the Panama Canal Watershed, resulting in the creation of five national parks (38% of the Watershed). This resulted in the most important achievement: the protection of the upper watershed basin that provides water for Canal operations and human use (water and electricity).

In 1997, after several studies were completed by the Regional Inter-Oceanic Authority (ARI), Law 21 was passed by Congress decreeing land zoning and protected land uses for the entire Watershed. Law 21 is not yet implemented, in fact, most of the inhabitants of the Watershed have no knowledge of Law 21, or its implications on their *modus operandi*.

The obvious question is where to start, what is indispensable? It is readily apparent, however, that the continued pace of development, without an integral watershed concept in mind, will affect the watershed's utility, evidenced

[Haworth co-indexing entry note]: "Foreword." Endara, Mirei E.. Co-published simultaneously in *Journal of Sustainable Forestry* (Food Products Press, an imprint of The Haworth Press, Inc.) Vol. 8, No. 3/4, 1999, pp. xiii-xiv; and: *Protecting Watershed Areas: Case of the Panama Canal* (ed: Mark S. Ashton, Jennifer L. O'Hara, and Robert D. Hauff) Food Products Press, an imprint of The Haworth Press, Inc., 1999, pp. xi-xii. Single or multiple copies of this article are available for a fee from The Haworth Document Delivery Service [1-800-342-9678, 9:00 a.m. - 5:00 p.m. (EST). E-mail address: getinfo@ haworthpressinc.com].

through indicators such as erosion, poor water quality, and increasing population pressure on the watershed's resources. The Panama Canal Watershed needs clear actions and committed actors who are aware of this region's complexity and can guarantee its future functioning.

The challenges ahead are exciting and unique. This group of bright Yale students has provided their insights on important issues, from policy-making to the impacts of exotic species. All are integral actions in the achievement of successful watershed management; their efforts are well-valued.

I take this opportunity to say that since the Yale visit in March 1998, the National Institute of Natural Renewable Resources (INRENARE) has become the National Environmental Authority (ANAM), with a stronger administrative capacity and integrity (environmental impact assessments, environmental standards, contamination issues, civil responsibility). We feel this institutional change will play a major role in promoting and enforcing a more integrated model of watershed management.

I would like to thank Mark Ashton, Jennifer O'Hara, and Taka Hagiwara, who are responsible for accepting the opportunity to participate in this challenge. Their visionary attitudes not only provided an excellent experience for Yale students, but for Panama too. Much remains to be done, and I invite all who are interested in participating in this challenge to join in making the opportunities ahead, a reality.

Mirei E. Endara
General Administrator
ANAM
Yale School of Forestry and Environmental Studies 1994

Preface

Within the last several decades, the challenges facing forest reserves and other protected areas around the globe have become increasingly complex. We have discovered that single resource management is no longer adequate and have turned to more holistic management by managing ecosystems or watersheds. We have learned that it was naïve to believe that boundary lines drawn on a map would result in protection on-the-ground.

Since the creation of Yellowstone National Park in 1872, natural resource managers have moved away from a strict preservation model to one that accounts for the procurement of goods and services necessary for human survival. There have been many other changes. For example, there is a trend towards privatization of protected areas. In many countries protected areas are being managed, if not owned, by private non-government organizations. Once banished from national parks after their homelands had been taken from them, there are now instances where local people remain as part of the protected landscape. Yet despite all of these positive changes, or perhaps as a consequence of them, the job of a protected areas manager has become notably more complicated.

We have learned, often the hard way, that protected areas management requires a clear understanding of the historical and social context that has shaped the circumstances particular to each site. For example, local consultation and participation are of paramount importance when creating a protected area or when establishing new land use regulations. Key stakeholders and managing entities should all have a place at the negotiating table. To ignore these key players is to place the stability of a protected area at risk.

Larger national and international factors must also be considered. For instance, tourism projects have been sometimes overly promoted in many protected areas as a means to generate local income. Despite advantages, there are many disadvantages as well. Much of the income generated by

[Haworth co-indexing entry note]: "Preface." Ashton, Mark S., Jennifer L. O'Hara, and Robert D. Hauff. Co-published simultaneously in *Journal of Sustainable Forestry* (Food Products Press, an imprint of The Haworth Press, Inc.) Vol. 8, No. 3/4, 1999, pp. xv-xvii; and: *Protecting Watershed Areas: Case of the Panama Canal* (ed: Mark S. Ashton, Jennifer L. O'Hara, and Robert D. Hauff) Food Products Press, an imprint of The Haworth Press, Inc., 1999, pp. xiii-xv. Single or multiple copies of this article are available for a fee from The Haworth Document Delivery Service [1-800-342-9678, 9:00 a.m. - 5:00 p.m. (EST). E-mail address: getinfo@haworthpressinc.com].

xiii

tourism leaves the country in the form of airfares and travel packages, while only a small proportion remains on-site. Furthermore, tourism is subject to a potentially fickle international market.

Despite the plethora of catchy slogans and acronyms (i.e., integrated conservation and development programs, sustainable development and use, etc.) the conceptual ideal of creating win-win situations through conservation and development is never as easily implemented. These goals of conservation and development are often at odds. Despite their familiarity with these goals, managers are often faced with unclear directives and an absence of priorities.

The purpose of this collection of papers is to examine some of the most current topics in protected areas management through a case study: the Panama Canal Watershed Area (PCWA). The challenges facing the PCWA are representative of those facing forest reserves and other protected areas around the globe, making it an excellent case for examination. Developing sustainable land-use systems in the Watershed is paramount to its future success. The primary challenge is to secure goods and services from the region without threatening the overall integrity of the ecological systems, while recognizing human needs and values.

This dynamic region contains an array of management issues that are sure to challenge even the most savvy natural resource manager. Within the last decade, demands on the Watershed Area have been increasing for water yield, hydro-power, economic development, recreation, conservation of biodiversity, pastureland, agriculture, and human settlement. Approximately 30 different agencies are responsible for administering programs and managing property within this area that spans 30 counties in five districts. Add to this complexity the final transfer of the Panama Canal and its associated real estate from the US Military to Panama by December 31, 1999.

This special issue includes eleven papers, the result of a graduate seminar entitled "Conservation for Biodiversity and Productivity" at Yale University, School of Forestry and Environmental Studies. This seminar was organized by ourselves (Ashton and O'Hara) in collaboration with Dr. Timothy Clark, a Policy scientist experienced in conservation of protected areas. During the spring of 1998, master's students gained insights into protected areas management from seminar lectures, library research, key informants, and a field trip to the Panama Canal Watershed Area, Panama. This combination yielded papers ranging in topics from agroforestry and invasive species, to policy processes and public participation. It is with a diversity of perspectives and a variety of tools that challenges in protected areas management will be met during the next one hundred years. We are confident that by examining models such as the Panama Canal Watershed Area, protected areas management will continue to benefit and evolve well into the next century.

ACKNOWLEDGMENTS

We thank Professor Graeme Berlyn for the gracious opportunity to publish these papers as a special issue of the *Journal of Sustainable Forestry*. Thanks to our sponsor the Yale School of Forestry for providing financial support for the Panamanian Field Trip. We are greatly indebted to our hosts, Director Mirei Endara and Rosa Cortez-Hinestroza of the National Institute of Natural Renewable Resources (INRENARE), whose generous assistance and superb coordination made our ten-day field trip to the Canal Watershed Area a most productive and memorable experience. Likewise to our host Taka Hagiwara of the Japan International Cooperation Agency, thank you for all of your logistical support. We are also grateful to our host Ira Rubinoff, Director of the Smithsonian Tropical Research Institute. We would have been lost without our trusty and knowledgeable guide Mario Bernal Greco. Thanks go to a host of presenters who took time away from their busy schedules to share with us the complexities of their work in the Watershed. They include: Stanley Heckadon Moreno, Roberto Ibanez, Tony Coates, Rick Condit and Egbert Lee, all from the Smithsonian Tropical Research Institute (STRI); Alfredo Mejia from the Inter-Oceanic Authority; George Hanily from Fundacion Natura; Mirei Endara, Erasmo Vallester, Indra Candanedo, Elizabeth de Laval, and Tomas Ramos of the National Institute of Natural Renewable Resources; Professors Mireya Correa and Noris Salazar from the University of Panama; Elisa Piti of the Association for the Conservation of Nature (ANCON), and Richard Moreno from the National Institute of Aqueducts and Sewers (IDAAN). Please forgive us any ommisions.

Mark S. Ashton
Associate Professor of Silviculture
Yale University
School of Forestry and Environmental Studies
New Haven, CT

Jennifer L. O'Hara
Doctoral Candidate
Yale University
School of Forestry and Environmental Studies
New Haven, CT

Robert D. Hauff
Research Ecologist
USDA Forest Service Institute of Pacific Islands Forestry
Kosrae State
Federated States of Micronesia

Introduction:
The Panama Canal Watershed Area

Jennifer L. O'Hara

TROPICAL WATERSHED MANAGEMENT

The Panama Canal Watershed Area (PCWA) represents perhaps one of the world's most complex managed ecosystems. The abundance of premium services provided by the Watershed's natural resources and the myriad of institutions involved in the oversight of this region make it a classic case for sustainable natural resources management. How best to manage this biologically diverse area is the subject of this volume.

In accordance with the Carter-Torrijos Treaty signed in 1977, the US military will transfer the Panama Canal and its associated real estate to the Republic of Panama on December 31, 1999. Understandably, Panama, a country of approximately 2.6 million inhabitants, is grappling with the overwhelming opportunity and responsibility that this land reversion holds.

The Canal has played not only a paramount role in the history of Panama and the United States, but globally as well. Since completion of the Canal in 1914, it has had far-reaching effects on world economic and commercial development by providing a shorter and more inexpensive passage between these two bodies of water. For example, a ship laden with bananas that originates in Ecuador and sails to Europe saves approximately 5,000 miles in transit (Panama Canal Commission 1994). With over 700,000 vessels using this service since its completion, this "short-cut" has not gone unnoticed (Panama Canal Commission 1994).

Jennifer L. O'Hara is a Doctoral Candidate, Yale School of Forestry and Environmental Studies, New Haven, CT 06511.

[Haworth co-indexing entry note]: "Introduction: The Panama Canal Watershed Area." O'Hara, Jennifer L. Co-published simultaneously in *Journal of Sustainable Forestry* (Food Products Press, an imprint of The Haworth Press, Inc.) Vol. 8, No. 3/4, 1999, pp. 1-9; and: *Protecting Watershed Areas: Case of the Panama Canal* (ed: Mark S. Ashton, Jennifer L. O'Hara, and Robert D. Hauff) Food Products Press, an imprint of The Haworth Press, Inc., 1999, pp. 1-9. Single or multiple copies of this article are available for a fee from The Haworth Document Delivery Service [1-800-342-9678, 9:00 a.m. - 5:00 p.m. (EST). E-mail address: getinfo@haworthpressinc.com].

Services

The Chagres River yields approximately 2.8 billion gallons of fresh water daily. From this total, 1.8 billion gallons (64%) are used by vessels to facilitate passage through the gravity-fed locks. This means that roughly 52 million gallons of water are necessary for each ship's passage. The second largest use of the water resource (896,000 gallons, 32%) is dedicated to the production of hydroelectric power. Finally, 6% of the daily yield (200 million gallons) supplies fresh drinking water for the nearby cities of Panama and Colón (Heckadon 1993).

Beneficiaries of Canal operations include international users of the Canal, the 7,500 permanent Canal employees, the people of Panama who derive approximately 6% of their gross domestic product from its operations, and the 1.5 million inhabitants of Panama City and Colón who depend on the PCWA for their potable water (USAID 1997). Additionally, the Watershed and the Inter-Oceanic Region support approximately 80% of the country's GDP and 50% of the country's population lives within this area.

A Feat in Engineering

From a technological point of view, the Canal is an engineering triumph that involved digging through the Continental Divide, constructing the largest earthen dam ever built up to that time, and designing and building the most massive canal locks and gates ever created. Today, this technological expertise continues to be demonstrated with the hydrological management of water supplied to Alhajuela and Gatun Lakes for Canal operation. Yet despite these advances, the Canal remains vulnerable to natural perturbations such as El Niño. The disruptions in weather patterns caused by El Niño in 1998 have decreased Gatun Lake's water storage supply to 81.5 feet, three feet below the optimal level (La Prensa 1998). This is the greatest deficit in the history of Canal operation. These losses have resulted in restrictions in transit draft for vessels using the waterway. The last time draft restrictions were put in place was in 1983, also as a result of El Niño (Panama Canal Commission 1998). Not only do these meteorological events affect the economy and operations of the Canal, but also the welfare of Panama's largest metropolitan populations whose water supplies are dependent upon this source. Canal operations must continue to plan for such natural perturbations as it is highly likely that the region will face more El Niño events in the future.

BIOGEOPHYSICAL CHARACTERISTICS OF THE PANAMA CANAL WATERSHED AREA

The Panama Canal Watershed Area comprises approximately 326,000 hectares of land and contains the Panama Canal within it (Figure 1). The Canal

FIGURE 1

Panama Canal Watershed

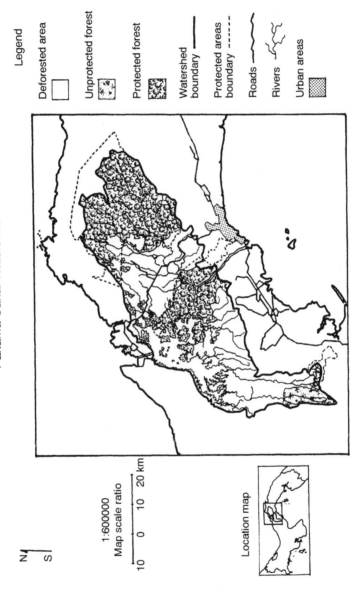

N
S

1:600000
Map scale ratio

10   0   10   20 km

Location map

Information taken from a satelite
image LANDSAT, sensor TM Band 3.54

Legend

Deforested area

Unprotected forest

Protected forest

Watershed
boundary

Protected areas
boundary

Roads

Rivers

Urban areas

bisects the Isthmus of Panama, stretching from Costa Rica to Colombia, and joins the Pacific and Atlantic Oceans. Fourteen percent of the Watershed Area is occupied by the artificial Lakes, Gatun and Alhajuela (also known as Madden Lake), which were created to meet the demands of Canal operation, metropolitan industrial and domestic water use, and hydroelectric power generation. The Madden Dam divides the watershed into the upper and lower basins (Figure 1). The lower basin consists of Gatun Lake and its main tributaries, the Gatun, Ciri Grande, and Trinidad rivers. The upper basin is composed of Lake Alhajuela and its tributaries: the Chagres, Pequeni, and Boqueron.

Climate in the watershed is characteristic of the lowland tropics and is classified as tropical monsoon according to the Köppen macroclimate classification system (Porter 1973). This climate type has a short dry season, however, rainfall during the wet season is usually sufficient to compensate for this period. All months receive over 6 cm precipitation (Porter 1973). According to meteorological stations within PCWA, the PCWA receives from 2000-3000 mm in annual precipitation (Houseal 1985).

The Watershed Area is dominated by lower montane wet forest as classified by the Holdridge Life Zone system (Porter 1973). Here *Ficus, Socreatea, Virola, Prioria* spp. among others, are common (Porter 1973). In forested areas, there are stands of old undisturbed forest, however, secondary forest ranging from 60-70 years in age is more common. Elevation ranges from sea level to approximately 300 m, the highest point being in Altos de Campana National Park. Winds are predominantly from the north and northeast and temperatures range from 23°-30°C (Graham 1985).

CONCERNS AND CHALLENGES

In 1952, 85% of the Watershed was forested. By 1983, this number declined to 30% as a result of forest land conversion to pasture and agricultural use (Heckadon 1993). From 1950-1980, populations in the PCWA increased by 80,000 people with 66% settling around the Lake Alhajuela area, a key water supply lake (Heckadon 1993). The resulting deforestation has led to increased erosion and sedimentation in the Canal Watershed. Heckadon (1993) estimated that during the period 1970-1985, Lake Alhajuela lost 5% of its water storage capacity.

Beginning in 1966 with the creation of Altos de Camapana National Park, the government of Panama created a series of protected areas in order to address the negative consequences of deforestation. Other protected areas include the three national parks: Soberania, Camino de Cruces, Chagres; the Metropolitan Nature Park, and the Lake Gatun Recreation Area. Although a total of 231,000 hectares of land are now under protected status, only 115,000 hectares are designated as protected forest. For example, in Chagres National Park, 30% of the land area is privately held. This has been attributed to inadequate funds for land purchase as well as the government's recognition that protected areas cannot be effectively managed using a traditional lock-up-the-resources park model. These protected areas are also extremely valu-able due to their biological resources. For example, in Chagres National Park 1,185 species of plants have been recorded (130 of which are endemic) while the entire area has yet to be inventoried (TNC 1995).

The Canal is surrounded by a ten-mile wide and 50 mile long forested belt, known as the Inter-Oceanic Region. This corridor has remained relatively intact as a result of the creation of the national parks mentioned above and management of lands by the US military. Outside the Inter-Oceanic Region and despite the number of protected areas, deforestation is occurring due to colonization pressures, large scale conversion of forestland to pasture and agricultural land, and industrial uses. For example, within the Watershed Area, the Mineral Resources Institute has permitted 22,000 ha of land to be mined for manganese, sand, limestone, and gravel while little monitoring is occurring (TNC 1995).

COMPLEXITY

The multitude of protected areas located within the Watershed Area is but just one example of the complexity of this Area. The protected areas within the PCWA span 30 different counties, in five districts, within the Provinces of Panama and Colon. Furthermore, 30-odd public and private agencies are responsible for natural resource management within the PCWA. Often these organizations have conflicting mandates and are unaware of each other's activities. Coordinating and integrating management activities represents a major challenge for this region (see Kahn, this issue). Finally, the Watershed represents a biologically complex and fragile ecosystem. Our understanding of tropical forest systems is only in its nascent stages.

METHODS

The following papers are the result of a semester-long graduate seminar on protected areas management and a eight-day field trip to Panama in March 1998, led by Drs. Mark Ashton and Tim Clark of Yale University. The seminar, entitled "Forest Conservation for Biodiversity and Productivity" ana-lyzed current trends in protected areas management through examination of a case study: the Panama Canal Watershed Area.

Field Observations

We spent ten days visiting various protected areas, canal and water treatment operations, and meeting with key informants (Figure 1). Our first day began with an orientation to the PCWA with presentations by Stanley Heckadon Moreno and Roberto Ibanez of the Smithsonian Tropical Research Institute (STRI). Heckadon briefed us on the Watershed's history and biological aspects. Next Ibanez introduced us to the Watershed's monitoring project, an interdisciplinary, multi-institutional endeavor that is monitoring forest cover, vertebrate populations, human population, hydrology and soils within the PCWA. The afternoon was spent observing Canal operations at Miraflores Locks.

On day two we visited the Agua Buena Agroforestry project bordering Soberania National Park and the Rio Cabuya Agrofrestry Farm Demonstration Project in the village of Cabuya de Chilibre (see Hauff's assessment of agroforestry projects). The afternoon was spent at the water treatment facility in Chilibre, a plant which provides drinking water for the inhabitants of Panama City. Here, we saw first-hand the effects of El Niño by observing the well-below normal water levels at Lake Alhajuela.

On the morning of day three, Dr. Antony Coates (STRI) gave us a trans-Isthmian tour of significant geologic formations. This was followed by presentations in the afternoon from Luis Alvarado, chief of the Operations Section, Meteorological and Hydrological Branch of the Panama Canal Commission and George Hanale from Panama's ecological trust fund, Fundacion Natura. Both representatives spoke about their agency's activities within the Watershed. Finally, Indra Candanedo discussed the role of protected areas managed by the Instituto Nacional de Recursos Naturales Renovables (INRENARE) in the PCWA.

Day four was spent in San Juan de Pequeni and La Bonga, communities located within Chagres National Park. This visit gave us the opportunity to observe land use patterns and have informal discussions with residents there.

Day five was spent in Soberania National Park meeting with education specialist Elizabeth de Laval, and head ranger Tomas Ramos, to discuss on-the-ground management issues with these INRENARE staff members.

Next we visited the east-side of the Watershed and met with INRENARE personnel to discuss forestry issues as well as agroforestry techniques.

The last day was spent in Campana National Park accompanied by Professors Mireya Correa and Noris Salazar of the University of Panama, Rosa Cortez of INRENARE, and an employee of TechnoServe. Here we learned about the botanical survey project being undertaken in the cloud forest by STRI and the University of Panama.

Summary of Papers

The articles contained within this volume are the result of informal interviews, information gathered from presentations by key informants, observations in the field, and library research.

The first set of papers focuses on the technical and biological aspects of management within the Watershed. Papers in the second section use a policy sciences framework to describe historical trends and social conditions in order to hypothesize future conditions and make recommendations.

The first paper by Cámara Cabrales analyzes trends in human migration into the Watershed region and describes the resulting deforestation. She discusses Panama's agricultural development policies as well as alternatives. The paper concludes with recommendations intended to reduce deforestation while simultaneously providing income for campesinos.

Next, Hammond discusses the problems presented by *Saccharum spontaneum* an invasive wild sugarcane species. Asian in origin, this vigorous plant has rendered thousands of hectares of forest and cropland in a state of useless arrested succession. Hammond describes the plant's physiology, its dominance in Panama, the current efforts to control it, and alternatives to these methods.

The next paper by Hauff is a qualitative assessment of three agroforestry projects located within the Canal Watershed. His paper elucidates criteria for evaluation, assesses the efficacy of three agroforestry projects, and concludes with recommendations.

The paper by Block explores the potential for carbon sequestration projects as a means to forest conservation and income generation for protected areas management within the Canal Watershed. Her paper reviews international efforts to develop carbon sequestration projects and provides data on carbon sequestration rates. She concludes with an evaluation of different types of carbon sequestration projects and highlights considerations.

Reversion of US military lands as it relates to protected areas management is discussed in paper number six. Corcoran's paper describes the responsibilities set forth by the 1979 treaty, available information on unexploded ordinances and toxic sites within the watershed, and the consequences of potential clean-up measures.

The next paper by Lowenberg provides information on sedimentation rates in the water supply reservoirs of the Panama Canal Watershed and discusses factors which influence these rates. Her paper compares methods for predicting sedimentation rates and discusses their application to the Panamanian Alhajuela and Gatun reservoirs.

The second section begins with a paper that examines the role of the Inter-Oceanic Regional Authority (ARI), an agency temporarily established to manage the allocation and development of lands within the Inter-Oceanic Region, an area formerly known as the Canal Zone. Whitney uses a policy sciences framework to assess the ability of the general and regional management plans, developed and managed by ARI, to protect natural resources within the Watershed.

Lieberknecht et al. provide a critical analysis of ecotourism for the Panama Canal Watershed. They draw upon expamples from other regions and give the benefits and disadvantages of encouraging such activities in Panama.

Next, a paper by Gardner discusses the quantity and quality of public participation in the Canal Watershed Area. His paper relates how organizations develop "myths" or perspectives that often lead to narrowly defined goals. Gardner describes the development of Panama's conservation movement, highlighting the role of the National Institute of Renewable Natural Resources (INRENARE).

Maxwell and Williams identify two conditions that contribute to the breakdown of the policy process as it relates to the PCWA. They provide an analysis of the obstacles to effective protected areas management and conclude by giving recommendations for improving the decision-making process.

The final paper by Kahn proposes that the lack of integration by the myriad of organizations operating within the Watershed is resulting in inefficient management of the PCWA's natural resources. He concludes by recommending prototyping exercises for building practiced-based experience, workshops for teaching integration skills and creation of an information coordinator who would be responsible promoting information flow between organizations.

Conclusions come in two forms. Firstly the paper by Clark and Ashton gives the protocol used by the class for assessing the resource issues of the Panama Canal Watershed. This paper provides a framework and context within which the students from the class worked, and gives the reader of the book a better understanding of some of the limitations and benefits of the analysis done by the students. Finally, Ashton and O'Hara provide a synthesis of the main resource issues concerning the Panama Canal Watershed.

CONCLUSION

At the broadest level, strategies to implement successful tropical watershed management must consider economic policies that control the use of tropical forests. Typically tropical forests have been undervalued for their rich endowment of services. Instead the benefits of timber extraction and conversion of forest to pasture and agricultural land are the primary foci. With the coming of the new millenium, the world will look towards the Panama Canal Watershed Area for answers. With Canal operation acknowledged as vitally dependent upon the Watershed's ecological integrity, the PCWA represents a case where human and ecological fates are inextricably tied and where a holistic approach to management is necessary. These conditions make it a fascinating and enlightening case for investigation. It is hoped that this special issue will shed light on the challenges faced by Panama and, on a broader scale, serve as an example for management around the world where people and landscapes are inseparable entities.

REFERENCES

Graham, A. 1985. Vegetation and vegetational history of Northern Latin America. Elsevier Scientific Publishing Company, New York.

Heckadon, S. 1993. The impact of development on the Panama Canal environment. *Journal of Inter-American Studies and World Affairs.* 35(3):129-149.

Houseal, B. 1985. Plan de manejo y desarrollo del parque nacional Soberania. USAID/INRENARE, Panama City, Panama.

Panama Canal Commission. 1994. The Panama Canal. Brochure published by the Panama Canal Commission, Panama City, Panama.

La Prensa. 24 March 1998. El Canal y la Danza de la Luvia. Article by Miren Gut. Panama City, Panama.

Panama Canal Commission. 1998. Panama Canal sets first draft restriction due to El Niño. February 12, 1998. *The Panama Canal Spillway,* Panama City, Panama.

Porter, D.M. 1973. The vegetation of Panama: a review. In: pp. 168-201, A. Graham (ed.). Vegetation and vegetation history of northern Latin America. Elsevier Scientific Publishing Company, Amsterdam.

TNC. 1995. Parks in Peril Source Book. The Nature Conservancy, Arlington, Virginia.

USAID. 1997. Activity Data Sheet. www.info.usaid.gov/pubs/cp97/countries/pa.htm

Small Farmer Migration
and the Agroforestry Alternative
in the Panama Canal Watershed

Luisa Cámara-Cabrales

SUMMARY. A combination of social, economic and physical factors contribute to a decision to migrate for a peasant farmer. Seventy percent of Panamanian land comprises steep topography with low soil fertility. Campesinos (peasant farmers) must perpetually migrate when the land they have settled in the forest loses fertility. This process of continual migration results in deforestation and conversion of forest land to pasture land, which then becomes degraded and subsequently abandoned. Also, government policy has not favored the activities of small land owners, but loans from national banks have financed the promotion of large-scale cattle ranches.

Migration patterns consist of moving to a forested area, clearing it for agriculture for one to three years, and thereafter sowing it for pasture. These conditions eventually result in low agricultural and cattle production. Campesinos therefore sell the land to cattle ranchers who can afford to invest in better management practices. Deforestation in the Canal Watershed, therefore, has left a pattern of land ownership comprising a matrix of a few large landowners, within which remain numerous but small peasant land holdings. A program for small land owners in and around the Watershed region that promotes social, cultural, and economic stability of land use is needed. Supporting and promoting agroforestry with uses of native tree species associated with

---

Luisa Cámara-Cabrales is a Doctoral Student, Department of Forestry and Wildlife Management, University of Massachusetts, Amherst, MA 01003.

[Haworth co-indexing entry note]: "Small Farmer Migration and the Agroforestry Alternative in the Panama Canal Watershed." Cámara-Cabrales, Luisa. Co-published simultaneously in *Journal of Sustainable Forestry* (Food Products Press, an imprint of The Haworth Press, Inc.) Vol. 8, No. 3/4, 1999, pp. 11-22; and: *Protecting Watershed Areas: Case of the Panama Canal* (ed: Mark S. Ashton, Jennifer L. O'Hara, and Robert D. Hauff) Food Products Press, an imprint of The Haworth Press, Inc., 1999, pp. 11-22. Single or multiple copies of this article are available for a fee from The Haworth Document Delivery Service [1-800-342-9678, 9:00 a.m. - 5:00 p.m. (EST). E-mail address: getinfo@haworthpressinc.com].

11

subsistence and market crops is an alternative for slowing deforestation and promoting sedentary land use. *[Article copies available for a fee from The Haworth Document Delivery Service: 1-800-342-9678. E-mail address: getinfo@haworthpressinc.com <Website: http://www.haworthpressinc.com>]*

KEYWORDS. Agroforestry, campesino, cattle, farmers, migration, native timber trees, peasants, ranching

INTRODUCTION

A combination of social, economic and physical factors contributes to the decision to migrate. Perhaps one of the most important physical aspects of Panama is that only 30% of the land is suitable for farming. Panama, out of all the Central American countries, has the greatest proportion of steep and dissected topography (Barry 1990). Cobos (1992) reported that 75% of the land can not sustainably raise livestock or produce annual agricultural crops. This physical aspect is further compounded by the socio-economic conditions in the country. After Brazil, Panama has the widest gap between rich and poor in Latin America (State of the World 1998). Campesinos must perpetually migrate when the land they have settled loses its fertility. This continual migration to different sites results in deforestation.

In this paper, I will analyze the migration patterns of campesinos into the Canal Watershed Area and other parts of Panama. In addition to analyzing issues concerning patterns of migration, I suggest changes in development policy that may provide alternatives to migration. Finally, I suggest that agroforestry systems that use native timber species with other crops may be an alternative to deforestation. Such systems can promote sedentary land use patterns for peasants, allow farmers to be self-sufficient, and prevent continued migration and forest clearance.

MIGRATION PATTERNS

Physical Characteristics in the Watershed

Elton (1997) reported that in Panama, 1.2 to 2 million ha of land with degraded soils now exist. In the Canal Watershed there is therefore a potential for sedimentation in lakes, particularly if forest cover is removed on steep topography (Barry 1990). Removal of forest may reduce the capacity of the Canal to operate at its current level (Alvarado 1998). The Watershed has an area of 326,225 ha, of which Gatun and Alhajuela Lakes cover 14% (Cortez 1989). Approximately 43% of the land is under protected area status, 35% is agricultural land, and 12.3% is human settlements (ANCON 1995). In general, 75 to 86% of the land is suited for forest and 16% for agriculture or cattle, but together more than 90% may be suited to a combination of forest and agriculture using appropriate agroforestry systems (Cortez 1989, ANCON 1995). It is important to mention that the 43% (182,400 ha) that is under protected area status is managed by the Institute of Renewable Natural Resources (INRENARE), thus demonstrating the importance of this institution in the forest conservation of the Canal Watershed.

Migration Patterns from the Interior

The typical campesino in Panama has two land use strategies: agriculture and cattle. Agriculture for subsistence crops includes corn (*Zea mays*), rice (*Oryza sativa*), beans (*Phaseolus vulgaris*) and cassava (*Manihot esculenta*); cattle are sold at the market to obtain cash for other foods and commodities such as cooking oil, detergents, radios and machetes. Interestingly, cattle are considered less risky investments since they are easy to transport on bad roads and there is a high national market demand. During interviews[1] with peasants and Embera indigenous communities living within Chagres National Park, they emphasized this issue. The villagers stated that it is not economically viable to take agricultural products to market as the roads are very bad, and transportation by river and bus are very expensive.

The "interior" includes Los Santos, Herrera, Veraguas, Chiriqui, and Cocle Provinces. All are located in the western part of the country on the Pacific coast side of the country (Figure 1). The peasant migrants, now in other regions of Panama, originally came from the "interior" of the country, but were displaced by cattle ranchers. The "interior" of Panama now has a high concentration of cattle lands owned by only a few ranchers (Cobos 1992).

Heckadon (1984) provides a case study for Los Santos, a province that is among those with high out-migration. With the independence of Panama from Colombia in 1903, the new government of Panama seized land in the province of Los Santos in the southwest part of the Watershed region and sold it to private landowners and companies. Many campesinos were therefore left without land. The government also promoted health services that reduced mortality, causing the population to double from 1920 to 1950. There was no alternative for campesinos but to move out of Los Santos Province.

By the 1930's practically all of the old land tenure systems from before 1904 had disintegrated (McKay 1984). Politics, along with low soil fertility, inefficient agricultural practices, and cattle management has therefore been an important driver for campesino migration.

FIGURE 1

Republic of Panama

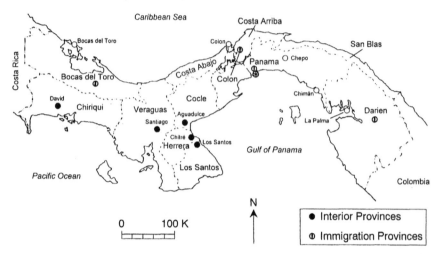

The "interior" has therefore served as an important source of migrants for the western and eastern parts of the Canal Watershed. The process of deforestation starts with forest clearance for cultivation of subsistence agricultural crops. This is practiced for one to three years and is followed by conversion to pasture land for cattle, which is then subsequently abandoned. When there is enough land available, the fallow period can last seven to ten years, allowing for the recovery of soil fertility. In this case a rotation of fallow permits the peasant family to subsist in the same area. The campesinos who convert all their land to pasture need more land to grow their basic staples. They must then ask other peasants with land in fallow for a plot to rent for cultivating. These are called medias or tercias[2] (Heckadon 1984, McKay 1984). If it is not possible to obtain medias, tercias, or rented land–a situation which is becoming common–the campesinos then try to find seasonal jobs in the nearest city or as either agricultural or livestock workers on the large ranches (Heckadon 1984, McKay 1984).

Campesino pasture land is often difficult and expensive to maintain because of vigorous weed invasion and costs of fertilization. Campesinos therefore sell the land to cattle ranchers who can invest in better livestock and pasture technology. Another alternative for the peasants is to ask cattle ranchers for cattle to pasture on their land. The arrangement consists of having the cattle for a period long enough to have offspring, giving back the original cattle to the rancher and dividing the young calves between the peasant and the rancher. When there is no more land available, the cycle of migration and forest clearing starts again, leaving a pattern of small land holdings belonging to the most successful campesinos, surrounded by land owned by a few large ranchers and agribusinesses.

This pattern of migration has extended to the frontiers of the provinces of Darien in the southeast and Bocas del Toro in the west (Figure 1) (Heckadon 1997). As land is being rapidly consumed, what will happen when no more land is available?

The percent of land for cattle has increased in part due to the demand for beef in the cities of Colon and Panama. Moreover, the government has promoted large-scale cattle ranching by providing credit. Cheap pasture for ranching has been acquired from peasants through government credit. As pasture land area increases, the area with forest cover decreases in the entire country. In 1950, there were 570,000 head of cattle and 550,000 ha of pasture; in 1970, there were 1.2 million head of cattle and 1.1 million ha of pasture (Heckadon 1984). Currently there are 1.4 million head of cattle and the forest area has been reduced to 44% of the original total forest cover of the country. Cobos (1992) estimated that by the year 2000 only 33% of Panama's forest will remain.

Canal Watershed Migration

The population in the metropolitan area (Panama City and Colon) is multi-ethnic, as a result of Canal construction that drew workers from around the country and abroad. In modern times, the attraction for jobs, services, and farmland contributes to high population and cultural diversity. The principal groups in the metropolitan area have come from Cocle, Chiriqui, Los Santos, Herrera, and Veraguas. Embera and Colombians also have arrived from the Choco Department of Colombia.

Outside the metropolitan areas the migration trends in the Canal Watershed have a similar pattern to that described for other rural parts of the country. Peasants from the "interior" have migrated to the marginal areas of the Canal Watershed to clear the forest and practice agriculture and cattle ranching. This has created a disorderly pattern of migration and land clearance in the Watershed with a social and economic marginalization resulting in poverty. Cortez (1989) interviewed migrants in the area and found a mean average land holding of 2.5 ha per family. Cortez (1989) refers to a USAID report of 1969 that stated that the small farmers are the largest socio-economic problem of Panama. It is estimated that 89% of the rural population lives under the poverty line. The same report stated that poverty is concentrated in the provinces of Veraguas, Panama, Chiriqui, and Colon. The rural population in those areas engages in subsistence agriculture. They do not have title to the land or credit to improve production and receive no technical advice, using only rudimentary technology. As recently as April 1998, *El Panamá América*, a daily newspaper, reported that 22% of the population lives in poverty, and of that percentage, 63% lives in rural areas.

Periods of Migration Within the Canal Watershed

In order to form Gatun Lake, the community of La Linea was flooded in 1912-1914. Before, people in this area performed subsistence agriculture. Following lake formation, they moved and worked for timber companies that extracted timber from forests around the lake (Cortez 1989). After timber extraction ended, the Gatun area was cleared for agriculture and cattle ranching.

On the western side of the Canal Watershed the province of Cocle is characterized by large land holdings. In Panama, the *interioranos*[3] of this region raise more cattle than any other group. Many of the peasants originally in Cocle migrated to work in banana plantations in Colón in the 1940's-1950's. In the 1960's, migration from the provinces of Los Santos, Herrera, Veraguas, and Chiriqui, resulted in the expansion of cattle ranches in the Watershed Area. These groups again settled on the west side of the Watershed in Arraijan, Chorrera, and Capira. In the 1960's and 70's the National Association for the Conservation of Nature (ANCON), the biggest environmental nongovernmental organization (NGO) in Panama, reported high deforestation rates affecting the region of Gatun Lake and the west sector of Alhajuela due to campesino migrants (ANCON 1995).

The last massive migration to the Canal Watershed started in 1979, in part due to the signing of the Panama Canal Treaty that called for a more collective use of the land within the Watershed (Cortez 1989). The Panama Canal Treaty stated that land under protection or used by the US army will revert back to Panama by the end of 1999.

ALTERNATIVES AT THE POLICY LEVEL

According to Heckadon (1997), poor development planning has caused the problem of deforestation in Panama. The small farmer crisis is a product of historic, ecological, and socio-economic circumstances.

The government's economic policy has promoted and financed cattle ranchers while neglecting subsistence agriculture. Data on cattle loans illustrates this: from 1970 to 1979, about $542 million was given in loans for cattle ranching, with an average of $54 million per year (Joly 1989). Data from the loans of the Panama Development Bank and National Bank of Panama reported about 90 to 98 percent of the total credit for farming was given to cattle ranching activity in the Watershed region (Heckadon 1984, McKay 1984, Heckadon et al. 1985, 1997).

The support for small landowners is insufficient and meager in comparison with the support for cattle ranchers. For example, the amount loaned by the government for cattle ranching is massive compared with a recent government announcement on March 26, 1998, that $39 million[4] will be used to combat rural poverty over the next five years. The Panamanian government needs to show a greater commitment to developing programs for small landowners in order to stop deforestation and alleviate poverty.

However, development planning should define short- and long-term programs. I summarize several important aspects to address rural development and conservation planning for the Canal Watershed. First, recognition of the necessity to adopt and promote sounder agricultural alternatives. Agroforestry systems are suited for sustainable farming production; the techniques used in current agriculture and livestock have to change (Heckadon 1997). Panama's government should diversify its economic program, focusing not only on the Canal Watersheds service-oriented economy, but on providing financial aid to promote small farmer programs inside and outside the Watershed.

Second, there is a need for a long-term program of market development for timber products obtained from the trees grown within the agroforestry systems, as well as for short-term planning for subsistence and cash crops. Currently, there is no strong market infrastructure for timber production and product processing. Most timbers are have either been imported as pre-processed lumber or have been exploited from the forest frontier. One avenue to explore to closely regulate timber exploitation might be forest product certification.

Third, a national program of land tenure rights and titles that restrict resale would promote more permanent settlements. This would provide peasants an incentive to adopt other farming techniques that can improve productivity. In Darien there is also a need for a strong program that will identify and delineate public land boundaries (such as parks), so reducing the incentive for colonization.

My fourth suggestion is that to succeed in the implementation and adaptation of agroforestry systems, subsidies and technical extension must be provided. INRENARE and non-governmental organizations are currently working to improve agroforestry systems, but they should increase the scale of their efforts. Similarly, and in conjunction with agrofroestry extension, education and public services (roads and communication) need to be provided to communities to promote stability and a sense of permanence.

Finally, coordination among government agencies and policy makers should be the rule and not the exception. INRENARE and the Ministry of Agriculture should have a coordinated agenda and a national objective of production with conservation of rural regions. This national agenda should emphasize conservation of soil fertility through combined cultivation of subsistence food crops, market crops, and tree cover.

ALTERNATIVES IN THE AGROFORESTRY PROGRAM

Agroforestry practices have been used by ancient civilizations of both tropical and temperate regions. Protection of soil from erosion by using agroforestry systems is well documented (Young 1989, Nair 1984). Agroforestry enhances soil by the addition of organic matter, prevents erosion by providing cover to the soil surface, and promotes soil water infiltration by improving soil porosity. For example, in the short term basic crops are harvested, in the middle term fruits and firewood are obtained, and in the long term there are timber products, thus lessening the farmer's risk in the market. Agroforestry systems make better use of resources by fostering diverse species and efficient use of nutrients. On soils with steep slopes, tree or shrub cover can help retain soil nutrients and prevent erosion (Nair 1984). Carefully designed agroforestry systems that match species combinations with the site in question can help campesinos settle the land more permanently and prevent migration into another forested area.

In many situations native species should be selected for planting in agroforestry systems (Table 1). Native species have adapted to the ecological and physical conditions of the soil and climate of the area. Also, native species have co-evolved with insects for pollination (Gentry 1987), and with fauna for seed dispersal (Leigh et al. 1992). I propose that agroforestry systems for the Canal Watershed should comprise site-specific mixtures of native tree species, basic agricultural food crops, and commercial fruit trees. The positive effects of agroforestry systems on the soil vary depending on the site's soil conditions, past land use history, species used, and the social and economic context of the farmers. Different methods should be used for each site depending on the farmer's desired product and income yield.

Examples in Panama of agroforestry systems can be found in Bocas del Toro and the village of Agua Buena in the Watershed region. In Bocas del Toro, CATIE and German Technical Aid (GTZ) have the following systems: *Theobroma cacao* (cocoa) associated with *Cordia alliodora* (timber species) with live fences using the genera *Tabebuia*, *Cedrela*, and *Bombacopsis* (Dixon and Dominguez 1993). The system of cocoa and *C. alliodora* in Costa Rica has also demonstrated high yield of product and maintained soil fertility (Young 1989).

Foresters in INRENARE and JICA (Japanese International Aid) are experimenting with mixtures of native and exotic species. Species being investigated for these purposes are: *Acacia mangium, Cedrela odorata, Swietenia*

TABLE 1. Native tree species with Latin and common names, by family, and by growth characteristics that make them potentially useful in agroforestry systems.

SPECIES, COMMON PANAMANIAN NAME	FAMILY	CHARACTERISTICS/USES
Anacardium excelsum (espave)	Anacardiaceae	Good timber species; shade tolerant; grows slowly at the first life stages and faster thereafter. General construction and local uses.
Tabebuia rosea (roble de sabana)	Bignonaceae	Hardwood, easy to work, good timber, fast growing and shade intolerant. Flooring, veneer, general construction and furniture.
Bombacopsis quinata (cedro espino)	Bombacaceae	Valuable timber, very promising for Panama, and has been used in some reforestation programs; easy to work and nail; resistant and durable to fungus. Veneer, general construction, furniture, plywood and pulp and paper products.
Cavanillesia platanifolia	Bombacaceae	Abundant species in Panama's forest; grows up to 40 m tall; considered a common persistant species in thin soils or dry ridges. It has local uses.
Swietenia macrophylla (caoba) and *Cedrela odorata* (cedro)	Meliaceae	Very valuable species; both have been over-harvested; considered from middle to shade intolerant species; existing broad market for timber. Veneer, interior decoration and funiture.
Prioria copaifera (cativo)	Leguminosae	Commercial hardwood with a very high use potential; recommended for upland and dry soils. Interior trim, furniture and cabinet work, veneer and plywood.
Calophylum longifolium (Maria)	Guttiferae	Considered good timber; local use in general constuction. Flooring, furniture, boat construction.
Terminalia amazonica	Combreretaceae	Flooring, furniture and cabinet work, railroad cross-ties, and general construction; used to build *piraguas* (boats) for the indigenous communities.
Dalbergia retusa (cocobolo)	Leguminosae	Potentially valuable species; used by indigenous Embera in the community of La Bonga for crafts. Handles, muscial and scientific instruments, jewelry boxes, and chessmen.
Spondias mombin (jobo)	Anacardiaceae	Edible for animals and humans; wood has local use and general carpentry.
Elaeis oleifera (corozo, Palma de mil usos)	Palmaceae	Many uses; frequently used by Panamanian peasant; all parts of plant are used.

TABLE 1 (continued)

SPECIES, COMMON PANAMANIAN NAME	FAMILY	CHARACTERISTICS/USES
Annona spraguie (chirimoya)	Annonaceae	Fruits consumed by farmers.
Anacardium occidentalis (Marañon)	Anacardiaceae	Fruits consumed by farmers; an exotic introduced long ago, but very well adapted and used by the farmers; used in Agroforestry in Panama
Cajanus bicolor (beans guandules, tree-bean)	Leguminosae/ Papilionoideae	Grows as a bush and is therefore an excellent crop to out-compete *Saccharum spontaneum*; used in agroforestry projects in Panama

References: Condit et al. 1993, Croat 1978, Chudnoff 1984.

macrophylla, *Bombacopsis quinatum*, *Cordia alliodora* and *Tabebuia rosea*. Mixing commercial forestry in a taungya system (cultivation of trees with crops) can be implemented, meeting the need for soil cover by preventing erosion while providing basic food crops and vegetables for harvest in the short term. Another alternative for commercial forestry is a soil cover using leguminous species such as *Pueraria phaseoloides* and *Centrosema pubescens*. At the ANCON Rio Cabuya Agroforestry Farm Demonstration Project, *Acacia mangium* was planted at 8 m × 8 m, with mahogany (*Swietenia macrophylla*) being interplanted one year later at about 12 m × 12 m. This was said to decrease the incidence of the shoot borer (*Hypsipila grandella*).

CONCLUSIONS

An adequate response to the migration crisis should move beyond technical alternatives for better soil uses that avoid erosion and deforestation. Technical solutions need to be accompanied by a national policy program, which contains social, cultural, economic, and ecological factors that will promote agroforestry programs for small and medium-sized farmers within the Watershed Area and others parts of Panama. I emphasize areas outside the Watershed just as much as inside because migrants come to the Watershed area looking for farmland after they have abandoned land elsewhere in the country (Heckadon 1984, McKay 1984, Heckadon et al. 1985, Heckadon 1997, and Joly 1989).

NOTES

1. Interview with peasants of San Juan Pequení and Embera indigenous of La Bonga in the Panama field trip of the course on Forest Conservation for Productivity and Diversity, March 17, 1998.

2. Medias is when the campesino without land plants and cultivates the land of another campesino, but the owner clears and cleans the land to start the cropping and they divide the harvest. Another type of Medias is when the campesino cultivates the land and gives it back to the owner with pasture planted. Tercias is when the owner of the land does not participate in any management and receives a third of the harvest Heckadon (1984).

3. Interioranos refers to people who came from the interior of the country, including mainly the provinces of Cocle, Chiriqui, Los Santos and Herrera.

4. Panama Presidency web site 1998. The $39 million to alleviate poverty and promote reforestation, rural agri-industries, and markets is part of a loan from the World Bank and the United Nations Environmental Program.

REFERENCES

Alvarado, K. L. A. 1998. Watershed hydrology of Panama Watershed. Talk at the Trade and Commerce School, March 16, 1998. Panamà.

ANCON, 1995. Evaluaciòn ecològica de la cuenca hidrològica del Canal de Panamà Report, Panamà.

Barry, T. 1990. Panama a Country guide. Inter-Hemispheric Education Resource Center. Albuquerque, NM.

Cobos Moran, J. A. 1992. Los recursos naturales renovable de Panamà. INRENARE. Panamà.

Condit, R., S. P. Hubbell, and R. B. Foster. 1993. Identifying fast-growing native trees from the Neotropics using data from large, permanent census plot. *Forest Ecology and Management*. 62:123-143.

Cortez-Hinestroza, R. M. 1989. Aproximaciòn Sociològica al Problema de la Migraciòn, Marginalidad y Pobreza del Campesino Asentado en la Cuenca Hidrografica del Canal de Panamà. Tesis de Licenciatura. Universidad Santa Maria La antigua. Facultad de Ciencias Sociales, Escuela de Sociologìa. Pp. 242. Panamà.

Chudnoff, M. 1984. Tropical Timbers of the World. USDA.

Croat, T. B. 1978. Flora of Barro Colorado Island. Standford University Press. Standfrod, CA.

Dixon, F. and L. Dominguez. 1993. El Proyecto agroforestal CATIE-GTZ en Bocas del Toro. In pp. 179-184. Heckadon Moreno S. (ed). Agenda ecològica y social para Bocas del Toro. Actas de los Seminarios tallers. INRENARE. Panamà.

Elton, C. (Coordinator). 1997. Panamà evaluaciòn de la sostenibilidad Nacional. CEASPA. Panamà.

Gentry, A. H. (ed.) 1987. Four Neotropical rainforests. Yale University Press, New Haven, CT.

Heckadon Moreno, S. 1984. La colonizaciòn campesina de bosques tropicales en Panamà. In pp. 17-44. Heckadon Moreno, S. and McKay, A. (eds). 1984. Colonizaciòn y destrucciòn de bosques en Panamà. Asociacion Panameña de Antropologia. Pp. 174. Panamà.

Heckadon Moreno, S. and E. J. Gonzalez, (eds.). 1985. Agonia de la Naturaleza. Ensayo sobre el costo ambiental del desarrollo en Panamà. IDIAP, Panamà.

Heckadon Moreno, S. 1997. Spanish rule, Independence, and the modern colonization frontiers. In Coates, A. G. (ed.). Central America: A Natural and Cultural History. Yale University Press. Pp 277. New Haven, CT.

INRENARE. 1995. Informe de Cobertura Boscosa 1992. pp. 37. Panamà.

Joly, L. G. 1989. The conversion of Rain Forest to pasture in Panama. In pp. 86-130. Schuman, D. A. and W. L. Partridge (eds). The human ecology of tropical land settlement in Latin America. Westview Press. Boulder, CO.

Leigh, E. G., A. S. Rand, and D. M. Windsor (eds). 1992. The Ecology of a tropical rain forest: seasonal rhythms and long-term changes. Smithsonian Institution, Washington, DC.

McKay, A. 1984. Colonizaciòn de tierras nuevas en Panamà. In pp. 44-62. Heckadon Moreno, S. and McKay, A. (eds). 1984. Colonizaciòn y destrucciòn de bosques en Panamà. Asociaciòn Panameña de Antropologìa. Panamà.

Nair, P. K. R. 1984. Soil Productivity Aspects of Agroforestry. ICRAF, Science and Practice of Agroforestry 1. Nairobi, Kenya.

Panama Presidencia de la Republica web site 1998.

Young, A. 1989. Agroforestry for Soil Conservation. ICRAF, Nairobi, Kenya.

Saccharum spontaneum (Gramineae) in Panama: The Physiology and Ecology of Invasion

Bruce W. Hammond

SUMMARY. *Saccharum spontaneum* (L.), one of two species of wild sugarcane, is a highly variable species ranging broadly in its native Asia and competing vigorously on extreme and disturbance-prone sites. The species has also become invasive throughout the tropics on agricultural soils degraded by fire and overuse. In the Republic of Panama, *S. spontaneum* is locking vast stretches of deforested lands in an arrested succession monoculture with apparently little or no value for agriculture or native wildlife. Despite the extent of the problem, surprisingly little research has been undertaken on the species or its control. Several reforestation projects underway in Panama have proven effective in controlling the grass but operate at a small scale. However, control efforts used for other invasive species such as *Imperata cylindrica* and *Dicranopteris linearis* suggest lower cost approaches for facilitating natural forest regeneration. For example, mimicking the natural colonization of old fields by woody plants, islands of fruit-bearing trees could be planted to attract animal dispersers and gradually expand into the surrounding grasslands. On *S. spontaneum* fields adjacent to remnant forests, forest plantations can produce timber while also catalyzing the natural regeneration of native tree species in their understory. *[Article copies available for a fee from The Haworth Document Delivery Service: 1-800-342-9678. E-mail address: getinfo@haworthpressinc.com <Website: http://www.haworthpressinc.com>]*

Bruce W. Hammond completed a Master of Forest Science degree at the Yale School of Forestry and Environmental Studies. Currently he is Field Ecologist with the Nature Conservancy on Marthas Vineyard, MA.

[Haworth co-indexing entry note]: "*Saccharum spontaneum* (Gramineae) in Panama: The Physiology and Ecology of Invasion." Hammond, Bruce W. Co-published simultaneously in *Journal of Sustainable Forestry* (Food Products Press, an imprint of The Haworth Press, Inc.) Vol. 8, No. 3/4, 1999, pp. 23-38; and: *Protecting Watershed Areas: Case of the Panama Canal* (ed: Mark S. Ashton, Jennifer L. O'Hara, and Robert D. Hauff) Food Products Press, an imprint of The Haworth Press, Inc., 1999, pp. 23-38. Single or multiple copies of this article are available for a fee from The Haworth Document Delivery Service [1-800-342-9678, 9:00 a.m. - 5:00 p.m. (EST). E-mail address: getinfo@haworthpressinc.com].

KEYWORDS. *Saccharum spontaneum*, wild sugarcane, invasive species, reforestation, natural regeneration, arrested succession, deforestation, swidden agriculture

TAXONOMY AND ORIGIN

The genus *Saccharum* belongs to the family Gramineae and the tribe Andropogoneae. *S. spontaneum* is one of two species of wild sugarcane from which domesticated sugarcanes were bred (Purseglove 1972). It is a perennial grass, which grows in small tufted forms 35 cm high as well as large erect forms that can grow over 8 m tall (Purseglove 1972), and can be distinguished from cultivated species by the thinner canes and narrow panicle. *S. spontaneum*'s native range extends from eastern and northern Africa, through the Middle East, to India, China, Taiwan, Malaysia and New Guinea (Purseglove 1972). It is common in India, less common in Southeast Asia, and somewhat rare in Africa (Panje 1970) (Figure 1). Varieties of the grass pioneer a range of habitat types and exhibit great plasticity in form. In India, New Guinea and Vietnam, for example, it initiates primary succession on new alluvial deposits, where it thrives in conditions of high light and moisture availability. Other habitats include river banks and flood basins, wetlands and poorly drained sites, and the edges of fields and forests. The grass has even been found growing up a forest tree like a climber, in fresh water like an aquatic plant, and directly on rock (Panje 1970). Climatic tolerance is

FIGURE 1. Estimated native range of *Saccharum spontaneum* in Panama (based on Purseglove 1972, Panje 1970).

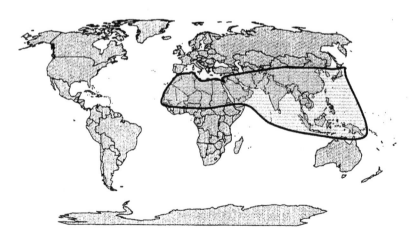

broad as well: the small, slender-stalked variety *S. indicum*, for example, survives winter temperatures of $-20°F$ (Alexander 1973).

Genetic analysis has revealed that *S. spontaneum* is a polyploid series. Forms with the smallest chromosome numbers are from northern India, which is probably the locus of origin (Purseglove 1972). The higher genetic variability of polyploid species is often associated with more variable morphology and physiology, enabling them to develop broad ranges and effectively compete in disturbance-prone and changeable habitats (Li et al. 1996). In many polyploid species, diploids specialize in wetter sites and polyploids in hotter, drier sites, but it is not known if this pattern holds in *S. spontaneum*. Despite this variability, identification of separate species within the *S. spontaneum* group has proved difficult and inconclusive. Continuous interspecific crossing within the genus only makes the taxonomy murkier. The *Saccharum* genus is also so closely related to the genera *Erianthus*, *Narenga*, and *Sclerostachya* that inter-generic crossing is possible, leading some botanists to place these species together in the "*Saccharum* complex" (Alexander 1973).

S. spontaneum is believed to have evolved in southern Asia, and during early Cretaceous times spread across what was then a continuous land mass extending to Australia. Isolation of the South Pacific islands caused a second species of wild cane to evolve there, *S. robustum*. This species was later bred into domesticated sugarcane (*S. officinarum*) and transported throughout the tropics (Alexander 1973). In this century, domesticated cane has been crossed with *S. spontaneum* to produce commercial hybrids that are more vigorous, hardy, and resistant to most of the major diseases (Purseglove 1972).

PHYSIOLOGY OF SUGARCANE

Sugarcane is the original C_4 plant–the C_4 photosynthetic pathway was discovered in experiments with sugarcane in the 1940's. In this type of photosynthesis, a four-carbon acid acts as a "CO_2 pump" to spatially separate the fixation of carbon dioxide from the leaf stomata, reducing the rate of photorespiration (the wasteful conversion of oxygen into carbon dioxide) (Crawley 1997). Because photorespiration increases at higher temperatures, C_4 plants have a competitive advantage in hot, sunny environments. They also have an advantage in drought conditions, because the more efficient photosynthesis process means their stomata do not have to be open as long and water loss is reduced. Sugarcane is considered an "extreme" C_4 plant as it can achieve extraordinarily high rates of photosynthesis at high light intensity and very low atmospheric CO_2 levels (Evans 1975).

Other mechanisms contribute to drought tolerance. During the early stages of water stress, leaf wilting reduces solar exposure. During the later stages of a drought, some varieties maintain tight stomatal control to virtually eliminate water loss and maintain a green canopy, while in other varieties the leaf canopy senesces during the drought and then regenerates with moisture (Evans 1975).

Sugarcane grows rapidly and maintains its hold on the canopy for an extended period. Three to five months after planting, sugarcane achieves a peak number of stalks. Although half these stalks die by nine months, the self-thinning has no impact on leaf area index (LAI). A LAI of three to seven only starts to decline about 16 months after planting (Evans 1975). The rapid early growth allows sugarcane to overtop other weedy competitors and eliminate them through prolonged shading. Studies show early leaf expansion is closely related to temperature (22°C being optimal) while later growth is more related to radiation levels (Evans 1975).

But sugarcane's most important competitive advantage is its diverse reproductive abilities. Sugarcane can reproduce by three different methods: layering, vegetative growth, and seeding. Canes have root primordia at their joints, which sprout when a piece of cane (called a sett in domesticated species) comes in contact with damp soil. Sugarcane expands aggressively underground by sending out rhizomes, tillers (secondary shoots), and tertiary shoots, which together form stools or clumps. The tillering pattern is so intricate and dense in *S. spontaneum* that it can almost be considered a sod-forming grass (rather than a tufted grass, which describes the rest of the genus) (Van Dillewijn 1952). The dense root mat creates intense root competition for any competing vegetation and is almost impenetrable for young seedlings. The rate of tillering is closely linked with light and temperature, and is most rapid at 30°C.

Certain environmental conditions trigger sugarcane to reduce vegetative growth and produce terminal inflorescences up to 25 cm long. Day length of 12.5 hours and night time temperatures of 20-25°C induce flowering, although different *S. spontaneum* varieties have different triggers (Evans 1975). Pollen is wind dispersed and remains viable for only a few hours (Purseglove 1972), although there may also be various degrees of apomixis (Panje 1970). Sugarcane seeds, which are actually fruits, are tiny (1 mm) and equipped with whorls of silky hairs for effective wind and water dispersal. The seeds have brief viability and germinate better in the light (Purseglove 1972).

S. SPONTANEUM AS AN INVASIVE SPECIES

In India and other Asian countries, varieties of *S. spontaneum* have rendered millions of hectares of croplands unproductive (Panje 1970). The species has been introduced or has escaped in Central and South America, Puerto

Rico, Florida and Hawaii. In all areas, it typically becomes invasive on poorly drained, poorly developed soils that have been subject to overgrazing and fire (Dinerstein 1979). According to one source, *S. spontaneum* was introduced to Panama in 1970 from southeast Asia. A ship crossing the Panama Canal was carrying construction equipment that had just been used to build an airstrip in Thailand for the U.S. military. Rain washed fragments of *S. spontaneum* into the Canal (J. Wright, pers. comm.). The species has spread across the entire isthmus–from Darien to Chiriqui–except for heavily forested areas (A. Taylor, pers. comm.). In the Panama Canal Watershed, the grass appears to dominate most lands that have been subject to slash-and-burn agriculture and are not under current cultivation (Figure 2).

REVIEW OF RESEARCH EFFORTS
ON SIMILAR INVASIVES

Because little research has been conducted on *S. spontaneum* as a weed species, this paper will review other invasive species problems in the tropics that may provide insights into the problem in Panama. All of the species considered create an arrested succession that inhibits forest regeneration, and the results of various reforestation methods are summarized in Table 1.

Imperata cylindrica (called *alang-alang* in Southeast Asia) is another Asian perennial pioneer grass that has spread to all continents. The genera *Imperata* and *Saccharum* are so closely related, in fact, that taxonomists put them both in the subtribe Saccharininae (Alexander 1973). Like *S. spontaneum, I. cylindrica* is highly variable, tolerates a broad range of soil types and moisture levels, is shade-intolerant, spreads vigorously via rhizomes, and produces many small wind-dispersed seeds.

In northeast India, *I. cylindrica* and a variety of other weedy pioneers are the first to colonize abandoned agricultural plots. During a 10 to 20 year fallow period, relay floristics occurs with a variety of annual, perennial, C_3 and C_4 weeds dominating for the first five years and then being shaded out by pioneer trees (Ramakrishnan 1992). In this process, *I. cylindrica* plays a valuable role in retaining nutrients and reducing erosion, while the presence of effective competitors prevents it from arresting the successional process. When the fallow period is shortened to five years, however, frequent burning, high erosion, and nutrient depletion leads to an arrested succession dominated by *I. cylindrica* (Ramakrishnan 1992). Repeated fire favors this grass because it can resprout from buried rhizomes, and regular burning turns the rhizomes into a dense mat that is almost impenetrable for other vegetation (Ivens 1983). Its C_4 photosynthesis provides an advantage in the heat and moisture stress of burn sites, and increases its nutrient use efficiency on eroded and over-cropped soils. Fire and a short fallow also builds up the

FIGURE 2. A landscape dominated by *Saccharum spontaneum* in the Panama Canal Watershed. Photo credit: H. Bradley Kahn.

weed representation in the soil seed bank and gradually eliminates tree seeds (Garwood 1989).

Significant research has been undertaken on the reforestation of grasslands dominated by *I. cylindrica*. Otsamo et al. (1997) tested 83 different tree species in Indonesian grasslands that were plowed, disked and weeded but not treated with herbicides or fertilizers. They found that several exotic species (*Acacia* in particular) rapidly formed a canopy dense enough to outcompete *I. cylindrica*. Native species survived but grew much more slow-

TABLE 1. Review of methods tested for breaking arrested succession, by author, location, species, results, and comments.

Researcher	Location	Invasive species	Methods tested	Results	Comments
Ashton et al. 1997b	Sri Lanka	*Dicranopteris linearis*	Late-successional native seedlings planted under intact/removed Caribbean pine plantation	Seedlings grew best when 3 rows of pines removed	Growth rates under pines and in openings varied by seedling species
Ashton et al. 1997a	Sri Lanka	Mixed grasslands	Cultivation, planting legumes, herbivory protection	Relative performance of legumes varied by site	
National Association for the Conservation of Nature (unpublished)	Panama	*S. spontaneum*	Cleaning grass, planting with cassava, maize, tree beans and teak	Grass controlled, crops produced, teak established	Cassava, maize, tree beans removed after 1 year
Cohen et al. 1995	Sri Lanka	*Dicranopteris linearis*	Cleaning ferns; cleaning and root removal; cleaning, root removal and tilling	Root removal produced highest diversity of natural regeneration	All regeneration from buried seed bank and seed rain; species composition dominated by herbs and shrubs
Communidad de la Union de Agua Buena (unpublished)	Panama	*S. spontaneum*	Cleaning of grass, planting tree beans and cassava, underplanting with native trees	Grass controlled, fruits and crops produced	
Guariguata et al. 1995	Costa Rica	Mixed fields	Evaluated natural regeneration under forest plantations	Plantation species with quickest canopy closure produced most regeneration	Herbaceous cover primary factor impeding natural regeneration
Nepstad et al. 1991	Brazil	Mixed grass/shrub fields	Planted scattered fruit-producing trees to attract dispersers	Trees grew well, not known if tree islands will become forest	Weeding and cultivation around seedlings produced good growth

TABLE 1 (continued)

Researcher	Location	Invasive species	Methods tested	Results	Comments
Otsamo et al. 1996	Indonesia	Imperata cylindrica	Plowing, planting exotic trees, underplanting with natives	Survival rates varied among native species planted	Recommend drought and heat tolerant native species for underplanting
Otsamo et al. 1997	Indonesia	Imperata cylindrica	Cultivation, tree planting	Exotic species grew faster than natives, controlled grass more quickly	
Parrotta et al. 1997	Many tropical sites reviewed	Various	Tree plantations that facilitate natural forest regeneration	Most effective on moist sites close to remnant forest patches	Mixed broadleaf plantation attract more diverse dispersers, produce richer regeneration
Ramakrishnan 1992	India	Imperata cylindrica	Swidden system with 5, 10, 20 year fallows	5 year fallow leads to domination by invasive weeds; 10 year fallow ideal	Imperata cylindrica outcompeted during longer fallows
Scott et al. 1994	Peru	Mixed fields	Planting legumes	Relative performance of legumes varied by site	Legumes shaded out by regeneration from adjacent forest in 53 months
Tuvey 1996	Indonesia	Imperata cylindrica	Plowing, herbicide (Roundup), slow-release NPK fertilizer	Higher growth of Acacia with herbicide/fertilizer than without	

ly. The authors suggest that exotic plantations should later be underplanted with natives. In another study, Otsamo et al. (1996) plowed or roto-tilled Indonesian grasslands three times and then planted *Paraserianthes falcataria*, a fast growing exotic tree. Two years later, the *Paraserianthes* had formed a closed canopy and shade-tolerant native species were planted underneath. Survival rates varied with species, and the authors conclude native species most tolerant of hot, bright, and drought-prone conditions should be selected. Finally, Tuvey (1996) found that herbicide and slow-release fertilizer significantly improved the growth of *Acacia* on former grasslands. The fields were cultivated with a Savannah three-in-one plough, treated with Roundup, planted, and then treated with NPK fertilizer.

Rather than trying to eliminate *I. cylindrica*, some traditional farming communities actually incorporate it into their agricultural system. In Sumatra, the Bataks burn *alang-alang* grasslands at the beginning of each growing season. These fields are then used to graze cattle and, after manually uprooting the grass' rhizomes, plant crops. This system can continue for seven years followed by only a brief fallow period, which allows *I. cylindrica* to regain dominance over later-invading, less palatable grasses (Sherman 1980). The Banjarese began using a similar system only 40 years ago when forest land was no longer available for shifting agriculture. They work *I. cylindrica* fields for four to seven years in a row; when other weeds invade, they leave the area fallow until *I. cylindrica* regains its dominance (Dove 1986).

Dicranopteris linearis–Forest clearing and burning in wet tropical climates can often produce monocultures of *Dicranopteris linearis*, a fern whose dense 1-2 m frond canopy can achieve arrested succession. Lawrence (1994) found that germination of tree seeds was actually higher *under* the ferns thickets on landslides in Puerto Rico, but seedling growth was much lower due to a twelvefold reduction in photosynthetic photon flux density versus open sites. The ferns are only outcompeted when tree seedlings established simultaneously or in gaps and edges of thickets shade them out.

Cohen et al. (1995) tested whether alternative disturbance treatments of fernlands on abandoned agricultural fields in Sri Lanka could facilitate rain forest succession. Their treatments included: cleaning of above ground vegetation; cleaning and removal of fern roots to a depth of 5 cm; and a combination of cleaning, root removal, and tilling. Results showed that any type of soil disturbance facilitated the establishment of herbs, shrubs, and trees, with root removal producing the highest species diversity and richness. The regenerating vegetation was dominated by herbs and shrubs with pioneer tree species comprising only 1-2%. The authors attribute all regeneration to seed rain and the buried seed bank.

Arrested Grasslands of Mixed Species

Nepstad et al. (1991) studied abandoned agricultural fields in the Parago-minas region in eastern Amazonia, Brazil. The fields are composed of a variety of grass and shrub species of low forage value and burn regularly. The authors identified the following factors as contributing to the arrested succession: low tree propagule availability (largely because forest birds and bats do not disseminate seeds into the large open areas); seed predation by ants and rodents; seedling predation by ants; seasonal drought; and root competition from the grasses. As a low-cost reforestation technique, they recommend speeding the formation of "islands" in grasslands by planting scattered, fast-growing, fleshy fruit-producing trees. These trees will attract seed dispersers and gradually expand into the grasslands. Prior to planting, sites should be weeded in a 50 cm radius and saplings planted in a hole filled with loose soil and composted organic matter. Using this method, exotics such as cashew (*Anacardium occidentalis*) began fruiting one year after planting and native species such as Brazil nut (*Bertholletia excelsa*) and mahogany (*Swietenia macrophylla*) showed good growth and survival.

Ashton et al. (1997a) studied successionally arrested mixed grasslands in the central hills of Sri Lanka which were subject to frequent fire, grazing, and herbivory. The authors planted four species of nitrogen-fixing legumes by seed in plots where all vegetation had been uprooted and removed. Wire mesh was used to protect half of each plot from browsing. Results showed that different species grew best on different sites (probably due to varying levels of herbivory, soil acidity, and moisture) and underline the importance of site-specific species selection. A similar study by Scott et al. (1994) in the Peruvian Amazon also found large differences between leguminous species, but after 53 months all the legumes were being shaded out by natural tree regeneration from the adjacent forest. Guariguata et al. (1995) studied natural tree regeneration under plantations of native species established in old fields in Costa Rica. They found that herbaceous cover was the primary factor impacting seedling establishment; plantations that were monocultures of the species achieved the quickest canopy closure and showed the greatest diversity and growth of tree seedlings.

THE S. SPONTANEUM PROBLEM IN PANAMA

Climatic factors constrain the nutrient supply in tropical ecosystems by promoting high rates of decomposition, soil respiration, soil acidity, and nutrient volatilization (Jordan 1991). Tropical forest systems have evolved highly effective mechanisms to retain and recycle nutrients (Jordan 1991), but forest clearing and burning unleashes dramatic physical and biologic changes including major nutrient losses, loss of soil organic matter, increasing pH, and changes in the soil microbial community (Nye and Greenland 1965, Ramakrishnan 1992). Cultivation exacerbates many of these changes by increasing soil erosion, further reducing soil nutrients and carbon, modifying soil structure, and increasing pests and weeds (Nye and Greenland 1965). Sustainable swidden agriculture depends on a fallow period to restore the natural processes. In Panama and elsewhere, the swidden system has broken down because the fallow period between firings is too short, *S. spontaneum* arrests succession on the fallow, and too little forested area remains to achieve natural forest regeneration.

In northeast India, Ramakrishnan (1992) has concluded that a fallow period of five years or less degrades the system, while a fallow of ten years is ideal. The restoration of soil organic matter during the fallow is essential for restoring soil nutrient levels and structure (Nye and Greenland 1965), but in India, organic matter continues to decline as long as the fallow land is dominated by *I. cylindrica*. As is probably the case with *S. spontaneum*, *I. cylindrica* provides low litter inputs to the soil and promotes high rates of litter decomposition. Nye and Greenland (1965) conclude that a grass-dominated fallow is generally less effective than a forested fallow in restoring soil structure and porosity, protecting the soil from erosion and compaction, and building litter and humus.

These differences are exacerbated by the much higher flammability of grasslands than forests. Grasslands concentrate organic matter near the ground and promote warm, dry conditions that make it combustible. Tropical forest canopies create cooler, moister conditions, and trees can maintain their canopy much longer into the dry season because their roots extend up to 12 m down. (Most grass and shrub roots are concentrated in the top 1-2 m.) In eastern Amazonia, Nepstad et al. (1991) found that the forests transpired 300 mm of soil water extracted from below 1 m down, which converted 13% of the annual solar radiation to latent heat.

Degraded soils and frequent burning disrupt the competitive balance between pioneer species and create a *S. spontaneum* monoculture that is difficult for other species to enter. The longer a site has been exposed to an intense disturbance regime, the more its seed bank is dominated by the standing vegetation rather than seeds from previously occurring species (Garwood 1989). Similarly, as forested areas in the landscape shrink and clearings expand, the seed rain of forest species in the clearings declines (Nepstad et al. 1991). *S. spontaneum's* underground growth becomes denser the longer it occupies the site, creating severe root competition for any potential invaders. Competitive exclusion is intensified by the drought conditions maintained in grasslands discussed above.

CONCLUSION:
CONTROL OF S. SPONTANEUM IN PANAMA

Several projects are underway in Panama to reforest lands dominated by *S. spontaneum*. At the Communidad de la Union de Aguabuena in Soberania National Park, for example, the grass is cut by machete before planting native tree beans *(Cajanus cajan)* and cassava *(Manihot esculenta)*. The *S. spontaneum* is kept cut until the planted trees have grown enough to shade out the grass, and then a variety of native fruit-bearing trees are planted (Figure 3). At the Cabuya de Chilibre reforestation project sponsored by the National Association for the Conservation of Nature, grasslands are interplanted with cassava, maize, tree beans (all of which are removed after one year) and exotic tree species such as teak *(Tectona grandis)*. In a version of the taungya agroforestry system, the crop cultivation keeps the *S. spontaneum* under control until the tree canopy begins to close. These projects are effectively converting grasslands to young forests, but because of the costs and labor involved the impacted areas represent a small fraction of the lands dominated by *S. spontaneum*.

The above case studies reveal that there are no quick and cheap "silver bullets" for converting degraded lands dominated by invasive species. Nonetheless, the research suggests some possible approaches for speeding the reclamation of *S. spontaneum* lands in Panama. Several studies successfully facilitated natural forest regeneration by mechanically disturbing the invasive species, planting shade-producing pioneer species, and then waiting for natural forest regeneration to occur. This may not be a viable option in Panama on large fields of *S. spontaneum* that probably have a limited representation of tree species in their seed banks and seed rain. Nepstad et al. (1991) propose an interesting alternative for sites that have lost their natural regeneration potential. By creating fruit-bearing tree islands to attract dispersers and gradually expand into the shaded areas at their edges, the strategy mimics the natural process that eventually allows most arrested grasslands and shrublands to be invaded by trees. It is not clear whether tree islands will naturally expand into grasslands dominated by such a tough and versatile competitor as *S. spontaneum*. If necessary, the cleaning and root removal treatment tested by Cohen et al. (1995) could be applied to the grasslands to facilitate natural expansion of the islands.

The strategy needs to be executed cautiously. A variety of studies have demonstrated that deep plowing alone can lead to vigorous *S. spontaneum* regeneration by resprouting rhizomes or wind-blown seed (Panje 1970). Accordingly, site disturbance needs to be followed by mechanical root removal, herbicide (a wide range of compounds have been found effective, including TCA sodium salt), and/or planting of shade-producing vegetation. To facili-

FIGURE 3. Tree beans (*Cajanus cajan*) planted on a former *S. spontaneum* field at the Communidad de la Union de Agua Buena in Soberania National Park, Republic of Panama. The legumes produce enough shade to control the invasive grass and improve soil properties with litter inputs. Photo credit: Bruce W. Hammond.

tate natural forest regeneration, probably only a narrow band of grassland around forest islands or edges should be treated at a time.

Another relatively low-cost approach is using a tree plantation to catalyze natural forest regeneration in its understory. Research from across the tropics indicates that forest plantations, while yielding economic benefits in timber production, can moderate forest floor microsite conditions, improve soil properties, and attract seed-dispersing birds and bats (Parrotta et al. 1997). In some cases, regeneration of native species beneath the plantation then occurs in as little as three years (Parrotta 1995), eventually yielding a diverse native forest with limited management intervention other than cutting timber. However, the catalytic plantation approach is most effective when remnant forest patches are close enough to allow natural seed dispersal and site conditions are relatively moist. The best plantation species are fast-growing pioneers that are native or exotics with low potential to become invasive. While mixed broadleaf plantings produce the most diverse understory regeneration (Parrotta et al. 1997), they are vulnerable during the sapling stage to fires sweeping the grasslands. Ashton et al. (1997b) successfully demonstrated the use of a fire-hardy exotic conifer (*Pinus caribaea*) to shade out invasive ferns and serve as nurse plants for late-successional native species. Catalytic plantations could be tested in *S. spontaneum* grasslands bordering existing remnant forests, and if successful in stimulating regeneration then progressively extended further into the grasslands. Since the choice of plantation species and silvicultural methods is dependent upon site conditions, the native species available for regeneration, populations of animal dispersers and other factors, an adaptive management approach is recommended where several alternative plantation designs are tested.

REFERENCES

Alexander, A.G. 1973. Sugarcane physiology. Elsevier Scientific Publishing Company, Amsterdam.

Ashton, P.M.S., S.J. Samarasinghe, IA.U.N. Gunatilleke, and C.V.S. Gunatilleke. 1997a. Role of legumes in release of successionally arrested grasslands in the central hills of Sri Lanka. *Restoration Ecology* 5(1):36-43.

Ashton, P.M.S., S. Gamage, I.A.U.N. Gunatilleke and C.V.S. Gunatilleke. 1997b. Restoration of a Sri Lankan rainforest: using Caribbean pine (*Pinus caribaea*) as a nurse for establishing late-successional tree species. *Journal of Applied Ecology* 34:915-925.

Cohen, A. L., B.M.P. Singhakumara and P.M.S. Ashton. 1995. Releasing rain forest succession: a case study in the *Dicranopteris linearis* fernlands of Sri Lanka. *Restoration Ecology* 3(4):261-270.

Crawley, M. 1997. *Plant Ecology*. Blackwell Science Ltd., United Kingdom.

Dinerstein, E. 1979. An ecological survey of the Royal Karnalibardia Wildlife Reserve, Nepal. Part 1. Vegetation, modifying factors, and successional relationships. *Biological Conservation* 15(2):127-150.

Dove, M.R. 1986. Peasant versus government perception and use of the environment: a case study of Banjarese ecology and river basin development in south Kalimantan. *Journal of Southeast Asian Studies* 27(1):113-136.

Evans, L.T. 1975. Crop physiology. Cambridge University Press, Cambridge, UK.

Garwood, N.C. 1989. Tropical soil seeds banks: a review. In: pp. 149-204. M.A. Leck, V.T. Parker, and R.L. Simpson (eds). Ecology of soil seed banks. Academic Press, New York.

Guariguata, M.R., R. Rheingans, and F. Montagnini. 1995. Early woody invasion under tree plantations in Costa Rica: implications for forest restoration. *Restoration Ecology* 3(4):252-260.

Ivens, G.W. 1983. The natural control of *Imperata cylindrica*–Nigeria and northern Thailand. *Mountain Research and Development* 3(4):372-377.

Jordan, C.F. 1991. Nutrient cycling processes and tropical forest management. In: pp. 159-180. A. Gomez-Pompa, T.C. Whitmore, and M. Hadley (eds.). Rain forest regeneration and management. Man and the Biosphere Series, Volume 6. UNESCO, Paris, and The Parthenon Publishing Group, Carnforth, UK.

Lawrence, W.R. 1994. Effects of fern thickets on woodland development on landslides in Puerto Rico. *Journal of Vegetation Science* 5:525-532.

Li, W.-L., G.P. Berlyn, and P.M. Ashton. 1996. Polyploids and their structural and physiological characteristics relative to water deficit in *Betula papyrifera* (Betulaceae). *American Journal of Botany* 83(1):15-20.

Nepstad, D.C., C. Uhl and E.A.S. Serrao. 1991. Recuperation of a degraded Amazonian landscape: forest recovery and agricultural restoration. *Ambio* 20(6):248-255.

Nye, P.H. and D.J. Greenland. 1965. The soil under shifting cultivation. Technical Communication No. 51, Commonwealth Bureau of Soils, Harpenden. Commonwealth Agricultural Bureaux, Farnham Royal, England.

Otsamo, A., G. Adjers, T.S. Hadi, J. Kuusipalo, and R. Vuokko. 1997. Evaluation of reforestation potential of 83 tree species planted in *Imperata cylindrica* dominated grassland. *New Forests*:127-143.

Otsamo, R., G. Adjers, T.S. Hadi, J. Kuusipalo and A. Otsamo. 1996. Early performance of 12 shade tolerant tree species interplanted with *Paraserianthes falcataria* on *Imperata cylindrica* grassland. *Journal of Tropical Forest Science* 8(3): 381-394.

Panje, R.R. 1970. The evolution of a weed. PANS. 16(4):590-595.

Parrotta, J.A., J.W. Turnbull and N. Jones. 1997. Catalyzing native forest regeneration on degraded tropical lands. *Forest Ecology and Management* 99:1-7.

Parrotta, J.A. 1995. The influence of overstory composition on understory colonization by native species in plantations on a degraded tropical site. *Journal of Vegetation Science* 6:627-636.

Purseglove, J.W. 1972. Tropical Crops: Monocotyledons. Longman Scientific and Technical, New York.

Ramakrishnan, P.S. 1992. Shifting agriculture and sustainable development: an interdisciplinary study from northeast India. Man and the Biosphere Series, Vol. 10. UNESCO, Paris and The Parthenon Publishing Group, Carnforth, UK.

Scott, L., C. Palm, and C. Davey. 1994. Biomass and litter accumulation under managed and natural tropical fallows. *Forest Ecology and Management* 67:177-190.

Sherman, G.D. 1980. What "green desert"? The ecology of Batak grassland farming. *Indonesia* 29:113-148.

Taylor, Alberto S. Botany Department, University of Panama. Personal communication, May 31, 1998.

Tuvey, N.D. 1996. Growth at age 30 months of *Acacia* and *Eucalyptus* species planted in *Imperata* grasslands in Kalimantan Selatan, Indonesia. *Forest Ecology and Management* 82:185-195.

Van Dillewijn, C. 1952. Botany of Sugarcane. The Chronica Botanica Co., Waltham, MA.

Wright, Joseph. Smithsonian Tropical Research Institute. Personal communication. May 28, 1988.

A Case Study Assessment of Agroforestry: The Panama Canal Watershed

Robert D. Hauff

SUMMARY. Finding sustainable land-use systems within the Panama Canal Watershed will be necessary for future management by the Panamanian government. Agroforestry is a land-use option for small-scale farmers living within the Watershed that can help achieve the goals of both conservation and productivity. This case study qualitatively evaluates current agroforestry projects in the Canal Watershed using an analytical framework based on other evaluations of agroforestry systems in Central America. Designated criteria for the analysis include: management objectives, project life span, incentives, technology, economic feasibility, community involvement, and extension. These factors can present obstacles to wide-scale adoption of agroforestry systems by small-scale farmers, thus preventing the realization of associated benefits of agroforestry. The analysis of the three field sites visited in March 1998 is followed by recommendations for expanding agroforestry practices among farmers in the Watershed. *[Article copies available for a fee from The Haworth Document Delivery Service: 1-800-342-9678. E-mail address: getinfo@haworthpressinc.com <Website: http://www.haworthpressinc.com>]*

Robert D. Hauff recently completed a Master of Forestry degree at the Yale University School of Forestry and Environmental Studies, New Haven, CT 06511. Currently he is Research Scientist, USDA Forest Service Institute of Pacific Islands, P.O. Box 82, Kosrae State, Federated Sates of Micronesia, 96944.

The author would like to thank all of the people in the villages discussed in the paper for all of their friendly help and willingness to provide information. His instructors Mark Ashton, Jennifer O'Hara, and Tim Clark helped with formulating ideas for this paper and his classmates also supplied many helpful comments. Lastly, he would like to thank all of the people who organized their wonderful visit to Panama and all of the people in Panama who did so much to make their stay a pleasant one.

[Haworth co-indexing entry note]: "A Case Study Assessment of Agroforestry: The Panama Canal Watershed." Hauff, Robert D. Co-published simultaneously in *Journal of Sustainable Forestry* (Food Products Press, an imprint of The Haworth Press, Inc.) Vol. 8, No. 3/4, 1999, pp. 39-51; and: *Protecting Watershed Areas: Case of the Panama Canal* (ed: Mark S. Ashton, Jennifer L. O'Hara, and Robert D. Hauff) Food Products Press, an imprint of The Haworth Press, Inc., 1999, pp. 39-51. Single or multiple copies of this article are available for a fee from The Haworth Document Delivery Service [1-800-342-9678, 9:00 a.m. - 5:00 p.m. (EST). E-mail address: getinfo@haworthpressinc.com].

KEYWORDS. Panama Canal Watershed, agroforestry, land-use, small-scale farmers, development projects, reforestation, protected areas

INTRODUCTION

Within the Panama Canal Watershed Area, specifically within Chagres and Soberania National Parks, agroforestry is currently being promoted as a means of protecting this region and its services by creating or maintaining forest cover. These services include providing and regulating sediment-free water for the operation of the Canal, generation of electricity, drinking water provisions for Panama City, biodiversity, carbon sequestration, and a livelihood for those who live within the Watershed and rely on management of its resources. Agroforestry can aid in protecting these services by preventing soil erosion and protecting soil fertility, creating both structural and species diversity, restoring degraded areas, providing corridors for animal species, and by bolstering agricultural productivity.

The following is a qualitative assessment of three different case studies involving agroforestry I observed during a visit to Panama in March 1998. The sites include the villages of Agua Buena and La Bonga, and the Rio Cabuya Agroforestry Farm Demonstration Project. The objective of this paper is to assess these agroforestry systems using designated criteria and to elucidate the key factors determining their success or likelihood for future success. Attempts will be made to offer proscriptions for future adoption of agroforestry practices in the Watershed based on the assessment of these sites.

A working definition of agroforestry from the literature will facilitate a clear discussion of the topic. Many definitions of agroforestry have been presented by different people (Nair 1989). For the purpose of this case study, a broad definition is adopted in order to include all of the important variables when discussing the efficacy of land-use systems. Agroforestry can be described by the following:

- a collective name for land-use systems involving trees combined with crops and/or animals on the same unit of land;
- a combination of production of multiple outputs with protection of the resource base;
- emphasizing the use of indigenous, multipurpose trees and shrubs;
- suitable for low-input conditions and fragile environments;
- involving the interplay of socio-cultural values more than in most other land use systems;
- structurally and functionally more complex than monoculture (Nair 1991).

Nair further states that the type of technology and practical applicability of an agroforestry system depend to a large extent on existing local conditions, both ecological and socio-economic. For this reason, it is necessary to examine local factors before designing agroforestry projects and to recognize heterogeneity within regions, nations, and landscapes.

Criteria for Evaluation

Despite extensive resources which have been devoted to development of agroforestry systems in Central America, adoption by small-scale farmers has been very low due to inappropriate application (Utting 1993). Farmer adoption remains one of the largest challenges to utilizing agroforestry for increasing sustainability and productivity. Regardless of a system's technical or environmental merits, it will have no impact unless a significant rate of adoption occurs (Raintree 1990).

To assess the three agroforestry sites, the following criteria were used: *management objectives, project life span, incentives, technology, economic feasibility, community involvement*, and *extension*. These criteria were selected from various assessments made on agroforestry systems throughout Central America (Current et al. 1995; Endara 1994; Conway 1989; Tschinkel 1987).

Management Objectives–Delineating management objectives is indispensable when devising any land-use system. Agroforestry is the tool used to achieve the management objective, not the objective itself. A clear understanding of these objectives is necessary to determine where objectives are complementary or overlap. In the national parks within the Canal Watershed there is a mix of stakeholders and their management objectives vary. Reforesting degraded lands and reclaiming areas dominated by *Saccharum spontaneum* (L.), an invasive grass species, appears to be a top objective of the Institute for Renewable Natural Resources (INRENARE) and the Panamanian government. Reforestation will prevent erosion and sedimentation, and regulate water flow, thus protecting the operation of the Canal and the provision of clean water to Panama City. Another objective of INRENARE is to protect biodiversity, to ensure ecosystem integrity and promote tourism opportunities for the national economy. The Ministry of Agriculture's objective is to increase production of cash crops among farmers. Non-governmental organizations have a variety of objectives that often include improving living conditions for local people, as well as self-strengthening for institutional survival of the organization. Typically, the objectives of small-scale farmers are to provide subsistence and/or cash by increasing the productivity of their land or other resources and, frequently, to retain land ownership within the family.

Project Life Span–The diffusion of agroforestry technologies hoped for by agencies such as CATIE and USAID has been prevented by low farmer adoption rates (Utting 1993). However, it is not surprising that subsistence farmers who live under insecure economic conditions are hesitant to convert their production system, especially since it might be their only means for survival. Continuous small-scale adoption, taking place over a 5 to 10 year period, is a more appropriate objective than intensive tree planting programs which may put the farmer's livelihood at risk (Current et al. 1995). Agroforestry projects lasting only a few years are therefore less likely to achieve the desired long-term result of wide-scale farmer adoption.

Incentives–Promoting farmer adoption by the provision of incentives is a frequently used strategy. However, financial incentives have proven to be counterproductive by creating dependency of farmers on a short-term project that might not be economically sustainable. Incentives can also be abused by farmers. Food-for-work incentives, where farmers are provided with food in exchange for working on community agroforestry projects, have also been found to be unsuccessful (Endara 1994). The minimal provision of material inputs such as tools or seedlings, however, is instead recommended to promote farmer adoption (Tschinkel 1987; Current et al. 1995).

Technology–The appropriateness of the technology transferred also affects farmer adoption. Low capital input and simple technologies are necessary for small farmers to participate. When traditional agroforestry systems already exist, such as swidden agriculture, efforts should be made to integrate existing knowledge and technologies (Utting 1993). Use of fast growing, site-insensitive species is extremely important in encouraging farmer adoption (Tshinkel 1987). However, over-emphasis of this strategy threatens biodiversity through the establishment of extensive plantations of exotic species such as teak (*Tectona grandis*). Less is known about fast growing native species, although efforts are being made to acquire ecological information on some of the faster growing natives to promote their use as reforestation species in Panama (Condit et al. 1993). Appropriate species selection with consideration to site variability and cultural preferences is essential for success.

Economic Feasibility–Economic feasibility is a requirement for long-term adoption by farmers. Relative profitability needs to exist, but risk avoidance through diversification can be a determining factor from the farmer's perspective (Arnold 1987). Although farmers usually adopt agroforestry systems to meet household subsistence needs, markets also play a significant role. Often local markets need to be developed for tree products in order for the systems to remain economically feasible. This is a problem in Panama where the wood product market is not well developed, especially for native species. Access to markets may not exist in remote areas thereby limiting profitability. A long-term vision is necessary since tree products have a longer production cycle than agricultural crops. Furthermore, tenure arrangements must be understood so that the economic benefits will go to the intended recipients (Bruce and Noronha 1987).

Community Participation–Community involvement has become recognized as an essential element in the success of agroforestry systems due to its neglect in failed projects (Utting 1993). Throughout the different stages of a project–planning, implementation, and monitoring–communities should participate and contribute ideas and information to ensure that the communities' needs and desires are met through the project design (Conway 1989). Providing community members with technical training to operate as local extensionists should be accompanied with organizational and administrative skills as the life of agroforestry support projects tends to be short. Community participation on technical decisions such as site and species selection should be accompanied by participation in monitoring and evaluation of project success.

Extension–Extension is often a crucial element in promoting farmer adoption of agroforestry systems. Very personal and intensive extension with repeated follow-up visits has been found to be vital to farmer adoption (Tschinkel 1987). Selecting villagers to train in agroforestry methods so that they can convey information to other villages has also been met with success (Utting 1993). The use of demonstration plots can reduce the costs of extension while increasing effectiveness (Current et al. 1995). Lastly, cooperation between agricultural and forestry extension is necessary for consistency in assistance.

DESCRIPTION OF CASE STUDIES

Agua Buena

Agua Buena is a small community (53 households) located on the border of Soberania National Park. It is connected to the Panama highway system with a paved road and public bus service. *S. spontaneum* dominates the hilly landscape making it prone to fire and preventing natural forest succession. In the past, the area had been used for hunting and collecting firewood. It was burnt in retaliation by the hunters who were prohibited from using the park for hunting, leading to the invasion of *S. spontaneum*. Conversations with farmers indicated that farmers perceive cultivation in areas dominated by *S. spontaneum* as impossible.

The INRENARE agroforestry project in Agua Buena began in 1992 and according to INRENARE is the most successful of 14 similar community projects around Soberania National Park. Participants plant and protect tree species while cultivating a variety of crops on the same plot for household and market consumption. This type of taungya system (see *Technology* below) has resulted in the establishment of trees (between three and six years old) in much of the eight hectare project area within the park.

Management Objectives–The participants' objectives are to provide food for their families and boost cash income with surplus production, while at the same time participating in an activity they enjoy. These farmers commute to Panama City for work, alleviating their dependence on the land for productivity. Adjacent to the park, their homes are furnished with electricity and plumbing, amenities not seen in more remote areas of Panama. INRENARE's objective is to protect the Watershed and its biodiversity by reclaiming sites dominated by the invasive grass through reforestation with native species.

Project Life Span–After reforestation has been achieved, the government-owned property will be taken out of agricultural production and the farmers will be offered land at other degraded sites where they will repeat the cycle. These sites will be located further away from the village than the original ones which are adjacent to the village.

Incentives–The incentives supplied by INRENARE for participating in the reforestation scheme have been limited to the use of government-owned land for agricultural production and minimal material inputs such as tools and seedlings. Although, after reforestation has been completed, the land will revert to INRENARE management as part of Soberania National Park. Community members will retain limited rights for harvesting fruit and other products with permission from INRENARE.

Technology–The technology utilized at Agua Buena consisted of a taungya system, whereby trees are cared for while inter-cropping annual crops, achieves the reforestation objective while providing participants with the incentive of crop production for consumption and the market. As the trees become established and crown cover begins to limit light availability to underlying crops, agricultural production ceases.

Timber species were planted at approximately 8 feet × 8 feet spacing and included natives such as *Ochroma pyramidale, Bombacopsis quinata, Guazuma ulmifolia, Anacardium occidentalis*, as well as the exotics *Acacia mangium* and *Gmelina arboraea*. The crest of the hill was widely planted with *Citrus* spp. and *Acacia mangium*, both of which are more tolerant to water stress than the native species. *Acacia* also improves soil nutrient status as a nitrogen-fixing legume. Fruit species such as *Anacardium excelsium, Mangifera indica* and *Bixa annato* were planted in the moister areas. Manioc, plantains, corn, papaya, and tree bean (*Cajanus cajan*) were the chief agricultural crops cultivated among trees or in patches where tree establishment had failed.

Economic Feasibility–The participants do not rely solely on the productivity of the land for income or subsistence. Risk avoidance is not a primary economic objective. Economic gain by selling surplus crops is facilitated by easy access to markets. Labor inputs were provided by family members.

Community Participation–Community participation in the project appeared to be minimal. Permission on what crops and trees can be planted must be granted by INRENARE. Although it was claimed that access to plots in this project is open to everybody, the peasants in a nearby village are not participating. This may be due to lack of community participation in project direction, as well as to the lack of incentives (no long-term tenure is available as participation concedes government ownership of the land). Additionally, if they do not have jobs in the city, economic insecurity could deter them from participating (farming on grass-dominated sites could be a risk).

Rio Cabuya

The Rio Cabuya Agroforestry Farm Demonstration Project was established in 1990 to provide a demonstration plot for farmers in the area, as well as to serve as an education center. Administered by the National Association for the Conservation of Nature (ANCON), a Panamanian non-governmental organization, this project promotes both production and restoration within communities in the area. Located on more than 64 hectares adjacent to Soberania National Park, it also serves as a forested buffer zone to the park. The site consists of mixed- and single-species plantations, an interpretive trail, a nursery, animal husbandry projects with local forest species, and a visitor pavilion. Funding is provided in part by the European Union and USAID. The influence of these agencies over operations at Rio Cabuya was unclear.

Management Objectives–ANCON's long-term objective is to provide educational assistance to communities in order to promote the incorporation of trees and native livestock into farmers' production systems and to provide environmental education to school children. Achieving financial self-sufficiency is an objective as well.

Extension–Seedlings can be purchased by farmers from the nursery at the site and information on cultivation practices is provided by ANCON foresters. ANCON requires demonstration of economic feasibility before it will participate in community forestry projects. Farmers are trained and encouraged to exchange their knowledge with other farmers.

Project Life Span–Indefinite.

Technology–The demonstration plots consist of plantations of teak (*Tectona grandis*), mixed acacia (*Acacia mangium*) and mahogany (*Swietenia macrophylla*), and spiny cedar (*Bombacopsis quinatum*), a native species with commercial value. The mixed plantations of acacia (8 feet × 8 feet spacing) and mahogany (12 feet × 12 feet spacing), with mahogany planted one year after the acacia, prevent damage to mahogany by the shoot borer, *Hypsipila grandella*. Predominant use of exotic species is attributed to fast growth rates as well as to existing markets for their timber. The closed canopy of the plantations prevents current cultivation of agricultural crops on much of the site. The nursery provides a diverse supply of seedlings including fruit trees and ornamentals.

La Bonga

La Bonga is a community located in Chagres National Park and is accessible by trail (approximately three hours by foot to the nearest road) or by boat on the San Juan de Pequeni River. It is located in an area where cattle pastures begin to give way to primary forest. La Bonga is populated by Embera people who migrated to this area from the province of Darien approximately 50 years ago. The village population is 288 (35 households) and has 600-1000 hectares for agricultural production. The community has usufruct rights to this specified area and practices traditional swidden agroforestry systems (Figure 1). Other economic activities include panning for gold, subsistence fishing, selling artisan crafts, and tourism.

Management Objectives–INRENARE's objectives for land-use in the park are to halt forest clearing (of both secondary and primary forest), reforest cleared areas, and incorporate trees into existing production systems. The

FIGURE 1. A swidden of the Embera people of the La Bonga community (population 288) within the boundaries of Charges National Park, Panama. Photo credit: Jennifer L. O'Hara.

objectives of the villagers are to raise agricultural production for consumption (rice was mentioned as being purchased, indicating that subsistence needs were not being met), as well as to raise cash income. Villagers expressed interest in acquiring credit for purchasing chickens and pigs or incentives for reforesting areas invaded by *S. spontaneum*.

Incentives–Although the villagers at La Bonga do not receive incentives for adopting sedentary agroforestry systems, INRENARE, as manager of the park, restricts villagers' use of the forest. Villagers are allowed to collect wood for fuel and construction from the surrounding forest. Trees for building dugout canoes can be extracted with permission from INRENARE.

Clearing for agriculture is only allowed in the specified 600-1000 hectare area and can only take place in forests of less than five years old. These restrictions effectively coerce the villagers to shift their traditional, land-extensive, swidden agriculture systems to sedentary, land-intensive, agricultural systems. Additionally, there is a disincentive for allowing fallows to reach more than five years of growth, preventing the recovery of cultivated land.

Technology–As previously mentioned, the villagers practice traditional swidden agriculture. Agroforestry systems observed in the village consisted of fruit gardens interspersed with cacao (*Theobroma cacao*), coffee (*Coffea arabica*), passion fruit (*Passiflora edulis*), peach palm (*Bactris gasipaes*), tree tomato (*Cyphomandra betaceae*), and plantains (*Musa* spp.). Manioc (*Manihot esculenta*), rice (*Oryza sativa*), and corn (*Zea mays*) were mentioned by villagers as being important crops. Coconut trees (*Cocos nucifera*) in home gardens were also prevalent in the village. One villager remarked during the visit that "the land is tired of cultivation," indicating exhaustion of the nutrient resources of their land under the current systems and a need for longer fallows.

Economic Feasibility–Agricultural production is solely for consumption with the exception of small amounts of coffee that are sold at the market. Market opportunities are restricted by the remote location of the village and volatile market prices. Cash income requirements are met by tourism, and to a lesser extent, by panning for gold. Villagers charge visiting tourists a US$7 fee per person. The villagers dress in traditional costume and perform dances for the tourists. Proceedings are equally distributed to participating households and supplemented by selling handicrafts. Additionally, panning for gold earns between US$.80 and 3.00 per day.

Community Participation–Community participation in determining park regulations and implementing forestry and agricultural extension was not evident. Furthermore, there is no management plan for Chagres National Park incorporating participation of those who live within the park boundaries.

Extension–In addition to promoting sedentary agricultural practices within the park, INRENARE encourages (the means by which are unknown to the author) use of subsistence instead of cash crops and use of leguminous species for green manure. Extension was not mentioned during the interview with villagers, although NGO involvement was perceived poorly in the village because of past experiences. Unfamiliarity with the Embera's culture and traditional agricultural practices could pose a barrier to successful extension.

ANALYSIS AND RECOMMENDATIONS

The three sites discussed above involve different aspects of agroforestry and are not meant for straightforward comparison and contrast, but rather to elucidate crucial and complex interrelating factors of using agroforestry as a tool for conservation and production. By clarifying the management objectives, I hope to present an evaluation of the projects' ability to accomplish them, as well as to discuss the contributing factors for success. Finally, suggestions for improvement of the current situation and recommendations for future projects are made.

Agua Buena, proclaimed by INRENARE as a success, owes much to its particular, and perhaps unique, factors. The excellent access and transportation allow the system to operate successfully. Participants are able to commute to Panama City and work for wages higher than those available in rural areas. Additionally, surplus production can be easily transported to the market, increasing profitability. Although INRENARE's termination of the project upon completion of reforestation is part of the original agreement with the farmers, conflict over use-rights might still arise. The fact that the participants do not rely on the site for subsistence makes INRENARE's termination of the project less difficult, albeit still problematic. The relocation of subsistence agriculturists' plots, however, could be potentially problematic.

Agroforestry as a biophysical tool to reclaim lands dominated by the invasive *S. spontaneum* has been demonstrated to be effective, and Agua Buena could serve as a model based on its technical merits. However, the economic elements of such a system might not be transferable to other social scenarios and broad application of this system is unlikely to meet success. Because of the low profitability of raising cash crops in remote settings, farmers do not have the incentive for adopting intensive production systems like those in Agua Buena. In areas around Soberania National Park, which is located near Panama City and largely accessible, this system has potential for replication. Finding participants who are interested in farming under INRENARE's stipulations, however, will be crucial.

Rio Cabuya effectively demonstrates technical feasibility of a variety of forest production systems. However, it seems to neglect the combination with agricultural crops involved in agroforestry systems, due in part to lack of suitable agricultural sites without crown cover. ANCON's objective to promote tree planting by farmers might better be achieved with demonstrations of agroforestry systems which supply short-term production (such as subsistence crops) as well as the long-term benefit of timber. The idea of using demonstration plots as an extension tool is certainly being embraced at Rio Cabuya, but seems to overemphasize forestry plantations which may not be acceptable to small-scale farmers because of the delayed financial benefits of such systems. Additionally, ANCON could be more active in promoting use of native, multiple-use species by incorporating such species into its demonstration plots as a pilot project. Although lack of markets is often cited as an impediment for using local species, unless a supply is created, the markets are unlikely to develop. ANCON is in a position to aid in this important market development process by generating supply of native timber, while relying on stocks of more valuable species for financial survival.

La Bonga is perhaps the most important case examined in this paper because it represents an increasingly common situation. La Bonga is a community in social transition due to land shortages, a situation seen throughout Panama. The villagers' traditional agricultural practices are no longer suitable under current conditions. At the same time they have entered into contact with the market economy and are seeking sources of cash income. Their management objectives–acquiring capital to produce more for the market–are at odds with INRENARE's conservation objectives, and the conflict is likely to heighten in the future. Adoption of sedentary agriculture is being accompanied by a reliance on the tourist industry for cash income. Agroforestry will continue to play an important role in the villagers' production system, but adoption of technologies desirable to INRENARE would be facilitated by understanding the traditional systems and working to adapt it to a system more acceptable to INRENARE. Because of the complicated social situation, community participation is especially crucial to finding management solutions acceptable to both parties. A management plan for Chagres National Park which incorporates community participation is needed (see Maxwell and Williams this issue).

One alternative is for INRENARE or a NGO to cooperate with the villagers in developing environmentally and culturally sensitive ecotourism opportunities while working to achieve sustainable and productive agroforestry systems. Concerted agricultural and forestry extension which strives to understand the community's needs and cultural traditions will be necessary in promoting appropriate agroforestry technologies. Promotion of multiple-use, nitrogen-fixing species, which are also native to the traditional lands of the Emberà such as *Inga* spp., by providing free seedlings and care instructions to interested farmers is one example of a possible extension strategy. Finally, improving market access for agroforestry products such as fruit, cacao, or coffee would help improve the profitability of land-intensive systems and meet the villagers' needs for increased cash income.

CONCLUSION

These case studies exemplify the complexity and diversity of agroforestry as a concept and in practice. Viewing agroforestry as a production system that incorporates both social and biophysical elements is necessary for reaching the specific management objectives. Recognizing the most important factors of a project's success or failure will facilitate the adoption of successful systems in difference locations. Each site will have a set of conditions that will require individually tailored systems. It is hoped that this analysis will create a better understanding of the role of agroforestry in the Panama Canal Watershed and how it can be used appropriately for increasing conservation and productivity.

REFERENCES

Arnold, J.E.M. 1987. Economic considerations in agroforestry. In pp. 173-190. H.A. Steppler and P.K.R. Nair (eds.). Agroforestry: Classification and management. Wiley, New York.

Bruce, J.W. and R. Noronha. 1987. Land tenure issues in the forestry and agroforestry contexts. In pp. 121-160. J.B. Raintree (ed.). Land, trees and tenure: Proceedings of an International Workshop on Tenure Issues in Agroforestry. ICRAF and The Land Tenure Center. Nairobi and Madison.

Condit, R., S.P. Hubbell, and R.B. Foster. 1993. Identifying fast-growing native trees from the Neotropics using data from a large, permanent census plot. *Forest Ecology and Management.* 62:123-143.

Conway, S.N. 1989. Assessing community participation in selected agroforestry projects. Cornell University Master Project.

Current, D., E. Lutz, and S. Scherr. 1995. Costs, benefits, and farmer adoption of agroforestry. World Bank Environment Paper Number 14. The World Bank, Washington, DC.

Endara, M.E. 1994. Socio-economic impacts of government agroforestry programs on farmers in two rural communities in Cocle Province, Panama. Tropical Resources Institute, Working Paper #69. Yale University School of Forestry and Environmental Studies, New Haven, USA.

Nair, P.K.R. 1989. Agroforestry defined. In pp. 13-18. P.K.R. Nair (ed.). Agroforestry systems in the tropics. Kluwer Academic Publishers, Dordrecht, the Netherlands.

Nair, P.K.R. 1991. State-of-the-art agroforestry systems. *Forest Ecology and Management.* 45:5-29.

Raintree, J.B. 1990. Theory and practice of agroforestry diagnosis and design. In: pp. 58-97. K. MacDicken and N. Vergara (eds.). Agroforestry: Classification and management. Wiley, New York.

Tschinkel, H. 1987. Tree planting by small farmers in upland watersheds: Experience in Central America. *The International Tree Crops Journal*. 4:249-268.

Utting, P. 1993. Trees, people, and power. Earthscan Publications, London.

The Potential
for Carbon Sequestration Projects
as a Mechanism for Conserving Forests
in the Panama Canal Watershed

Nadine Block

SUMMARY. Carbon sequestration projects are currently being explored as a method for offsetting carbon emissions and addressing global concerns over climate change. This paper examines the potential for carbon sequestration projects in the Panama Canal Watershed, an area with increasingly fragile forest ecosystems. The maintenance of forests in the Watershed has economic and ecological value for operating the Canal, providing drinking water to Panama City, and protecting wildlife habitat and diversity. Land conversion to agriculture and pasture has led to widespread deforestation; predicted population growth threatens the Watershed further. Compounding the problem is the lack of funding for protected areas in the Canal Watershed Area. Carbon sequestration projects are being promoted as an economically efficient way to reduce harmful greenhouse gases in the atmosphere and increase forest cover in developing countries. Forests, acting as sinks for carbon, can reduce emissions produced elsewhere. Many concerns have been

---

Nadine Block recently completed a Master of Forestry degree at the Yale School of Forestry and Environmental Studies and is currently employed at the Pinchot Institute, 1616 P Street NW, Washington, DC 22202.

The author would like to thank Dr. Mark Ashton and Jennifer O'Hara for their assistance in preparing this document. In addition, Daniel Shepherd, Manrique Rojas, Claire Corcoran, Sarah Whitney, Maya Loewenberg and Chris Williams provided valuable comments.

[Haworth co-indexing entry note]: "The Potential for Carbon Sequestration Projects as a Mechanism for Conserving Forests in the Panama Canal Watershed." Block, Nadine. Co-published simultaneously in *Journal of Sustainable Forestry* (Food Products Press, an imprint of The Haworth Press, Inc.) Vol. 8, No. 3/4, 1999, pp. 53-66; and: *Protecting Watershed Areas: Case of the Panama Canal* (ed: Mark S. Ashton, Jennifer L. O'Hara, and Robert D. Hauff) Food Products Press, an imprint of The Haworth Press, Inc., 1999, pp. 53-66. Single or multiple copies of this article are available for a fee from The Haworth Document Delivery Service [1-800-342-9678, 9:00 a.m. - 5:00 p.m. (EST). E-mail address: getinfo@haworthpressinc.com].

53

raised by developing countries, and uncertainties relating to carbon storage and carbon trading remain to be resolved. Despite these draw-backs, carbon sequestration projects offer an economically attractive strategy for furthering Panama's goals of protecting and expanding forest cover in the Watershed. *[Article copies available for a fee from The Haworth Document Delivery Service: 1-800-342-9678. E-mail address: getinfo@ haworthpressinc.com <Website: http://www.haworthpressinc.com>]*

KEYWORDS. Carbon sequestration, Panama Canal Watershed, forest management, sustainable forestry, protected areas, joint implementation

INTRODUCTION

The Panama Canal Watershed, encompassing approximately 325,000 hectares (Greenquist 1996), includes several protected forest areas. These protected areas provide numerous ecological and economic services, but involve both direct and indirect costs (Greenquist 1996, Heckadon 1993). There are the costs associated with operating and maintaining protected areas. There are also lost costs from not being able to use the land for forest harvesting, agriculture or development. A developing country such as Panama may find it difficult to justify the expenditures of protected areas management, as it is not perceived as promoting economic development (Heckadon 1993). One strategy for generating funding for protected areas in the Watershed is the development of carbon sequestration projects. Carbon sequestration projects involve international agreements in which units of carbon sequestered from the atmosphere in one location are traded to offset emissions elsewhere. Besides offering viable economic benefits to Panama, there are significant ecological benefits including a reduction in atmospheric carbon, maintenance and expansion of forest cover, retention of soils, and protection of biodiversity (Brown and Adger 1994).

This paper explores the potential for carbon sequestration projects in the Panama Canal Watershed. I review concerns related to protected areas, background on international efforts to develop carbon sequestration projects, and data on carbon sequestration. Finally, I evaluate the types of projects that might be appropriate in the Watershed and considerations that should be addressed.

PROTECTED AREAS IN THE PANAMA CANAL WATERSHED

The Canal Watershed

The importance of the Canal Watershed lies in its function as a natural system that provides a constant water supply for the Canal and for the residents of the Panama City metropolitan area (Heckadon 1993, INRENARE 1990). The clay soils of the Watershed erode easily and are damaged by the removal of protective vegetation. The soils are nutrient-poor and typical of those found in tropical forests, of which perhaps 3% may be suitable for agriculture (Heckadon 1993). Unfortunately, deforestation is occurring at a rapid pace and is threatening the fragility of the Watershed. It is estimated that over 180,000 hectares were deforested from the early 1950s to the late 1980s (Heckadon 1993). The main causes of deforestation are land conversion to agriculture and pasture, mainly due to population growth and migration (see Camara, this volume). Deforestation leads to erosion of the soil and contributes to the build-up of sediment in the Canal (Heckadon 1993).

Setting aside protected forest areas in the Watershed is considered by some to be the best strategy to combat deforestation and preserve the Canal (Greenquist 1996). Additionally, protected areas offer preservation of biodiversity, wildlife habitat, provision for ensuring high water quality, and opportunities for education and tourism, which have been listed as issues of concern to the Panamanian government. National parks are the most common form of protected areas in Panama, but other types of protected areas include biological preserves, forest reserves, and recreation areas.

Management and Financial Concerns

The management of Panama's protected areas is primarily the responsibility of the Institute for Renewable Natural Resources (INRENARE), created in 1986. Altos de Campana, the first national park in Panama, was created in 1966. Soberania, at 22,104 hectares, and Chagres, at 129,000 hectares, are the largest national parks in the Canal Watershed (INRENARE 1994). They were created in 1980 and 1984, respectively, primarily to ensure the conservation of the hydrographic basin of the Panama Canal.

Besides its own operating budget, INRENARE receives support from several Panamanian organizations. A trust fund was established in 1995 to ensure the long-term conservation of Panama's natural resources. Fundacion Natura, a Panamanian nongovernmental organization (NGO), administers grants from the fund, which was established with money from United States Agency for International Development (USAID), The Nature Conservancy, and the Panamanian government. Half of the annual interest from this fund, approximately $700,000 per year, goes directly to INRENARE for management of its protected areas (Hanily, pers. comm.; Endara, pers. comm.). In addition, the Institute receives limited assistance from the National Association for the Conservation of Nature (ANCON), a conservation NGO, in operating protected areas.

As is the case in many other countries, however, insufficient funds are allocated to the maintenance of Panama's protected areas. INRENARE has to stretch their operating budget to cover a variety of aspects of its protected areas, and has been exploring additional methods for generating financial support for protected areas. The Institute has explored options such as charging user fees for national parks and developing ecotourism. However, user fees have not been very successful because of the cost of collecting fees; ecotourism is still in the developing stages.

One strategy for generating support of protected areas that several developing countries have pursued in the last few years is carbon sequestration projects. The idea behind carbon sequestration projects stems from the growing concern over climate change and increasing interest in policies designed to mitigate emissions of greenhouse gases (GHGs). Carbon sequestration projects offer a potential opportunity to increase both financial and institutional support for reforestation and protected areas management.

POLITICAL AND TECHNICAL BACKGROUND ON CARBON SEQUESTRATION PROJECTS

Joint Implementation

There is increasing evidence that humans have had a significant impact on climate change through anthropogenic emissions of GHGs (IPCC 1996). In response, the United Nations Framework Convention on Climate Change (referred to as FCCC or the Climate Convention) was signed in Rio de Janeiro during the 1992 Earth Summit. The objective of the Climate Convention is to stabilize atmospheric concentrations of GHGs to prevent dangerous interference with the climate system (Figueres et al. 1996). The participants in the Convention discussed numerous measures to reduce GHG concentrations. One such measure is Joint Implementation (JI). JI is a concept that refers to international agreements through which an entity in one country partially meets its commitment to reduce GHG levels by offsetting its domestic emissions with projects it finances in another country (Figueres et al. 1996; USIJI 1996). This would take place only if there were a binding agreement on GHG limits (such as a carbon quota or tax).

The economic basis for JI is that it offers a least-cost scenario. Assuming international differences in marginal costs of response options and policies, JI is likely to reduce the overall costs of a given set of climate policies, increase the scope for climate action at a given cost level, or both (Andrasko et al. 1996). If country B (typically a developing country) could reduce GHG emissions more cheaply than country A (presumably a developed country), then A could basically buy those reductions in B by investing in a carbon offset project. From an ecological and economic standpoint, JI presents a "win/win" situation: country A gets partial credit for reducing GHGs and country B gets the investment and the ecological benefits of the development project (Zollinger and Dower 1996).

The global economic and environmental benefits of an international carbon trading system are potentially huge. Zollinger and Dower (1996) cite estimates that the global costs of meeting any given GHG reduction target could be reduced 50 to 70 percent B compared to the cost without international cooperation B through a trading program that took advantage of regional differences in the costs of implementing projects.

JI projects can take many different forms, but the two main types are land use projects and energy projects (Figueres et al. 1996, Cutright 1996). Land use projects include both carbon sequestration through practices that measurably increase the carbon fixing ability of a certain area of land, and preservation of natural carbon stocks (in soils and forests) threatened with imminent destruction. Sequestration refers to the natural process by which the carbon in gaseous compounds is incorporated into plant biomass (Cutright 1996). Examples of land use projects include forest preservation, reforestation, afforestation, and sustainable forest management projects. Energy projects include fuel switching, cogeneration or application of renewable energy or energy efficiency projects that reduce net fossil fuel GHG emissions (Figueres et al. 1996).

The cost of buying carbon credits is still being negotiated, and there are wide disparities. Average costs of sequestration through afforestation have ranged from $1 to $8 per ton of carbon in tropical countries (Brown and Adger 1994); in some cases, countries have been able to charge as much as $40 per ton. Costa Rica has sold over 1,000 permits at a cost of $3 per ton of carbon (Tangley 1998), and has recently sold "certified tradable offsets" at $10/ton (Goodman 1998). A carbon sequestration project in the Rio Bravo Conservation and Management Area of Belize is costing $1.91 per ton (PfB et al. 1994). When the pilot phase is over, and if binding agreements are made on JI projects, there is a general expectation that a carbon market will evolve into a more realistic supply and demand structure with higher carbon values and steeper investments (Figueres et al. 1996).

Concerns About JI

Multiple concerns about JI have been raised by developing countries and NGOs. One key issue is responsibility. Developing countries argue that since developed countries have produced the majority of GHGs, they should be responsible for reducing their emissions rather than buying their way out of the situation. Critics of JI insist that it allows industrialized countries to continue their relatively high rates of GHG emissions, while simply offsetting them in low-cost settings overseas (Zollinger and Dower 1996). Environmental NGOs worry that JI will reduce the incentive to develop more efficient technologies since developed countries can buy their way out of commitments to improve their own environmental performance (Zollinger and Dower 1996).

Developing countries also have concerns over sovereignty, worrying that they will lose control over what can be done within their own borders. The offsetting of one country's GHG emissions in another country involves an explicit exchange in property rights (Brown and Adger 1994). It has also been argued that there is an inevitable rising curve in the costs for carbon offsets. If JI permits developed countries to purchase all of the cheap credits at the onset, this would leave only the costlier options for developing countries when they have to reduce emissions themselves (Figueres et al. 1996).

Because of concerns about JI, a political compromise was worked out in 1995. A pilot phase was established during which projects would be referred to as "Activities Implemented Jointly" (AIJ). It was agreed that AIJ projects would meet certain criteria: consistency with national development priorities; endorsement by the governments of the participants; achievement of measurable emission reductions which would not have occurred without the activity ("additionality"); and financing that is additional to current official development assistance funding ("financial additionality") (Figueres et al. 1996, Cutright 1996). Examples of AIJ projects are summarized in Table 1.

TABLE 1. Examples of carbon sequestration projects implemented jointly, a brief description of their activities, their location, and reference.

Type of Project	Brief Description	Country	Source
Reforestation	Reforestation of 500 ha with teak to establish commercial plantation	Panama	USIJI 1997
Reforestation	Conversion of 2,000 ha of active hayfield to natural forests	Russia	USIJI 1996
Forest Conservation	Addition of 2,000 ha of tropical forest to the already established 2,000 ha Bilsa Reserve	Ecuador	USIJI 1996
Reduced Impact Logging	Implementation of harvesting guidelines aimed at reducing logging damage on 1,400 ha of commercial forest	Indonesia	Putz and Pinard 1996
Forest Conservation + Sustainable Forestry	Addition of 6,014 ha to existing Rio Bravo Conservation and Management Area (RBCMA); Development of sustainable forestry management on 47,379 ha of RBCMA	Belize	PfB 1994

Uncertainties and Challenges

Although several carbon offset pilot projects have been implemented under the AIJ trial period, the future of JI is very uncertain. Companies have been eager to launch JI projects with the anticipation that if and when carbon quotas are established, they will be in the forefront of the activity. It is likely that carbon prices in the future will be costlier, so companies that participate now are getting the best deals (Goodman 1998). There is also hope that companies may receive "early reduction credits" (LeBlanc 1997). However, JI projects in a voluntary system combine substantial risks with low financial returns (Zollinger and Dower 1996). Few countries have a carbon tax and no country has set carbon quotas, both of which are measures that would induce carbon trading. It is unlikely, given the atmosphere at the recent Kyoto Conference, that a system for carbon trading will be established earlier than 2008 (EDF 1998). Until a system is established, there is no monetary value for carbon offsets.

Because developing countries were reluctant to agree to a binding JI program at the Kyoto Conference, projects between developing and developed countries are uncertain. However, developing countries at the Kyoto Conference proposed a similar concept called the Clean Development Mechanism (CDM) (EDF 1998). The objectives are to allow the developed countries to meet part of their commitments of GHG reductions overseas, and to help the developing countries to achieve sustainable development and therefore contribute to the objective of the FCCC (Guardia, pers. comm.). Although still in the conceptual phase, the basis for CDM is that, similar to JI, projects could be financed by industrial nations as a cost-effective way to meet some of their emission-reduction obligations (EDF 1998). Although developed countries are enthusiastic about carbon offset projects, developing countries refused to make explicit commitments in Kyoto to participate in CDM.

Carbon Storage

Carbon sequestration rates and total amounts of carbon stored in tropical regions are highly variable and uncertain. The rate at which trees take up or sequester carbon is directly related to growth; the total amount of carbon stored in a tree depends on size or total biomass (Smith 1993). The amount of carbon in the forests is the total sum of carbon stored in the living biomass, dead biomass, and the soil (Burschel et al. 1993). A study of carbon stored in U.S. forest ecosystems reveals that the majority of the carbon is stored in soils (59%), while the remainder is in trees (31%), litter (9%), and understory (1%) (Trexler 1991). It is estimated that forests sequester more than 92% of the world's terrestrial carbon and store between 20-100 times more carbon per hectare than agriculture does (Cairns and Meganck 1994).

The range of carbon stored per hectare is dependent on different definitions of forest types, on assumptions of soil fluxes and characteristics, and on the previous land use of the forested area (Brown and Adger 1994). There can be significant differences between primary forests, secondary forests, and plantations in terms of carbon sequestration rates and total storage capabilities. Primary forests contain large amounts of carbon but exhibit little or no net growth or additional carbon storage; second-growth forests contain less carbon but continue to take up and store carbon (Smith 1993). Plantations can have tremendously high rates of growth and uptake. When plantations reach a mature stage, the net carbon sequestration stops. Therefore, the value of plantation forestry in carbon sequestration rests in its temporary utility.

Rates of aboveground carbon accumulation in tropical forests range widely between negative values (when stands are degrading) to higher than 15 Mg C/ha/year (1 Mg = 1 metric ton) in fast-growing plantations (Houghton et al. 1991). Young secondary forests (less than twenty years old) can accumulate carbon at rates of 3.5 to 5 Mg C/ha/year (Houghton et al. 1991). Similarly, carbon storage varies by forest type and geographical region (Table 2). Brown and Adger (1994) estimate that a closed primary tropical forest can store approximately 283 Mg C/ha and closed secondary forest can store 152 to 237 Mg C/ha.

TABLE 2. Amount of carbon storage (Mg C/ha) by forest type, region, and reference.

Forest Type	Region	Amount of Carbon Stored (Mg C/ha)	Reference
Tropical moist (aboveground biomass)	Brazilian Amazon	114-160	Brown and Lugo 1992
Primary diptocarp	East Malaysia	348	Putz and Pinard 1993
High-density mixed moist	Tropics	239-366	Brown 1997
Low-density mixed moist	Tropics	169-245	Brown 1997
Unproductive Forest	Panama	71.9	Hall and Uhlig 1991
Managed forest	Panama	88.8	Hall and Uhlig 1991
Forest left undisturbed for 60-80 years	Panama	113.3	Hall and Uhlig 1991
Closed primary	Tropics	283	Brown and Adger 1994
Closed secondary	Tropics	152-237	Brown and Adger 1994

To assess the potential for carbon sequestration by native species on tropical plantations in Panama, investigators studied the growth of five tree species native to the tropical moist forests of Panama (CTFS 1997). The study found that carbon sequestration rates varied widely, ranging from 0.38 Mg C/ha/yr for *Enterolobium cyclocarpum* (Fabaceae family) to 4.6 Mg C/ha/yr for *Hura crepitans* (Eurphorbiaceae family). Additional research on the carbon storage potential of plantations and primary forests would be useful in determining the effectiveness of carbon sequestration projects.

POTENTIAL FOR CARBON SEQUESTRATION PROJECTS IN THE PANAMA CANAL WATERSHED

Existing Support Mechanisms in Panama

Despite the uncertainty, there is tremendous potential for carbon sequestration projects, and it is worth exploring in the Panama Canal Watershed. The Panamanian government has already demonstrated support, both in theory and in practice, for carbon sequestration projects.

Although not specified as such, carbon sequestration projects would be compatible with INRENARE's 1990 National Plan for Forestry Action (INRENARE 1990). The plan outlines twenty proposed action programs grouped into five main categories: (1) programs designed to improve agricultural practices and promote sustainable agroforestry systems; (2) programs designed to introduce sustainable yield in forest management for timber harvesting as well as industrial and social reforestation projects; (3) programs that promote the sustainable use of firewood in urbanized areas; (4) programs that seek to conserve forestland by expanding national protected areas; and (5) programs designed to improve institutional cooperation among the public and private organizations interested in forest protection.

A recently approved General Environmental Law says that the State recognizes carbon sequestration as an environmental service of the forests and may establish the mechanisms to finance projects through joint implementation (Guardia, pers. comm.). Zollinger and Dower (1996) describe a proposed "Panama Climate Action Investment Fund" that would allow Panama to combine JI projects for financing purposes. Backed by a consortium of national and international conservation organizations (including The Nature Conservancy) in consultation with the Panamanian Ministries of Planning and Natural Resources, the proposal is seeking investments in a greenhouse fund. The proposal suggests that GHG offsets will be offered at a fixed price, investors will receive a small revenue-based return to be paid out after 20 years, and partial investment in 20-year U.S. Treasury Bonds will secure the fund against risk (Zollinger and Dower 1996). The implications of such a fund are uncertain, but it might entice monetary commitments from venture capital or "green funds," which can accept lower rates of return, though it would not generate major revenues in the short term (Zollinger and Dower 1996).

There is already one approved AIJ project in Panama, titled "Commercial Reforestation in the Chiriqui Province." The project involves the reforestation of degraded pastureland in the Chiriqui Province in western Panama. The organizations involved in this project are CAOBO (a Panamanian company providing implementation and financing), Fundacion Natura, and the U.S.-based Center for Clean Air Policy.

According to the project description, 500 hectares will be planted with teak (*Tectona grandis*), at a density of 1,235 trees/ha (USIJI 1997). After site preparation and planting, the land will be established and maintained as a certified teak plantation and managed in a sustainable manner ("sustainable" is not defined in the document) as a source of high quality hardwood (USIJI 1997). The site will be thinned every six years until the project lifetime (25 years) expires, at which time the trees will be harvested and processed. The non-commercial and slash debris will be sold to local people for firewood, thus replacing the current supply of firewood, which is primarily harvested in a non-sustainable way (USIJI 1997). It is expected that the plantation will be replanted at the end of the 25-year harvest cycle. The GHG benefits of the project are estimated to result in net sequestration of 15,720 tons of carbon, based on the estimated amount of woody biomass carbon that is harvested and stored indefinitely in durable wood products, minus the carbon that would be released through activities such as burning firewood and processing end products (USIJI 1997).

As with any JI project, there are costs, benefits, and concerns. The capital cost to purchase the land and establish, maintain, and operate the plantation for the first seven years is $3,700,000; after year seven, the plantation is expected to be self-sufficient through the sale of harvested wood (USIJI 1997). In addition to the carbon sequestration benefits, the report claims that the project will alleviate pressure on natural forests by providing an alternative source of fuel-wood and will create jobs outside of traditional agriculture. Furthermore, the project will assist the government in Panama in promoting reforestation of previously degraded lands. Concerns about the project include the vagueness of the term "sustainable" and uncertainty about what will happen to the land after the plantation is harvested.

Potential Types of Carbon Sequestration Projects in the Watershed

Carbon sequestration projects may include forest preservation, reforestation, afforestation, and sustainable forest management, all of which have advantages and disadvantages in the Panama Canal Watershed. Not all types of projects would be appropriate for protected areas, but would further the goal of maintaining forest cover in the Watershed.

It has been argued that preservation of primary forests for carbon sequestration may be more efficient and have fewer negative long-term ecological, social, and political results than plantations established for this purpose (Cairns and Meganck 1994). It may be less expensive to slow deforestation by protecting natural areas than to reforest large areas (Cairns and Meganck 1994). Furthermore, forest management that employs conservation of primary forests concurrently supports the goals of maintaining biological diversity.

Conversely, plantation forestry may be appropriate where there is an urgent need for watershed rehabilitation or where it can substitute for unsustainable industrial timber harvesting (Cairns and Meganck 1994). Unfortunately, many plantations in Panama consist of monocultures of exotic species such as *Eucalyptus*, *Acacia*, and *Gmelina* that are "biodiversity deserts" (Hubbell and Foster 1992). Research on the carbon storage potential of native species will be valuable to the development of carbon sequestration plantations comprised of native species.

Although the predominant approach recommended for carbon offset projects is planting trees, an ecologically sound and economically viable alternative is controlled timber harvesting to reduce the environmental damage and carbon release due to logging (Putz and Pinard 1993). Uncontrolled and unrestricted logging, common to developing countries, can be extremely detrimental to long-term ecological and economic productivity. Putz and Pinard (1993) point to numerous studies that show that logging damage can be substantially reduced through directional felling and planned extraction of timber on properly constructed and carefully utilized skid trails. The authors highlight numerous advantages of controlled selective logging over plantation forestry, including: immediate carbon benefits at low expense, maintenance of native species on site, lower fire risks, and lower likelihood of soil erosion or degradation.

Further Considerations

There are numerous opportunities for carbon sequestration projects in the Watershed that will generate funding for protected areas (such as forest preservation projects) or will contribute to economic sustainability (plantation or low-impact logging projects). However, several factors should be examined. One consideration is whether projects may be implemented in coordination with national parks. Due to the criteria of "additionality," which states that a project must prove that emission offsets would not have occurred were it not for the proposed activity, INRENARE would not be able to receive funding for carbon sequestration in an already protected forest.

One option, however, would be to consider a park expansion project, into an area that is threatened for deforestation in the near future. Another option is to increase a carbon sink within a park by reforesting a degraded area within its boundaries.

Another consideration is coordination with the Inter-Oceanic Region Authority (ARI). ARI is responsible for overseeing the reversion of U.S. lands in the 10-mile wide strip known as the Inter-Oceanic Region (formerly the "Canal Zone") to the Panamanian government. ARI is currently exploring options for development and forestry projects in the Inter-Oceanic Region. Therefore, this would be a prime opportunity to consider carbon offset projects in that area. INRENARE, Fundacion Natura, and ANCON might all be appropriate organizations to explore carbon offset partnerships.

CONCLUSION

There are still many unresolved issues regarding carbon sequestration projects. In particular, without a carbon tax or quota system currently in place to induce carbon trading, companies are investing in carbon sequestration projects without the assurance of future economic returns. Additionally, valid concerns among developing countries relating to responsibility and sovereignty render JI or CDM projects uncertain.

However, there are certainly benefits to be realized by such projects, particularly to a developing country like Panama that must protect the Canal Watershed. Maintenance and expansion of forest cover in the Watershed is crucial to the health of the Watershed, the operation of the Canal, and the provision of drinking water to Panama City. Threats to the Watershed in terms of land conversion are increasing, especially with the reversion of the Canal in 1999. Furthermore, INRENARE and other organizations charged with protecting the forest resources of the Watershed must find financial support for the services that they provide. Foreign investment for carbon sequestration projects would offer an economically attractive strategy for protecting the Watershed and should be investigated.

REFERENCES

Andrasko, K., L. Carter, and W. Van der Gaast. 1996. Technical Issues in JI/AIJ Projects: A Survey and Potential Responses. Background paper prepared for the Critical Issues Working Group, UNEP AIJ Conference, Costa Rica.

Brown, S. 1997. Estimating biomass and biomass change of tropical forests. FAO Forestry Paper #134. Food and Agriculture Organization of the United Nations. 55 pp.

Brown, K. and W.N. Adger. 1994. Economic and political feasibility of international carbon offsets. *Forest Ecology and Management* 68: 217-229.

Brown, S. and A.E. Lugo. 1992. Aboveground biomass estimates for tropical moist forests of the Brazilian Amazon. *Interciencia* 17(1): 8-18.

Burschel, P., E. Kursten, B.C. Larson, and M. Webber. 1993. Present role of German forests and forestry in the national carbon budget and options to its increase. *Water, Air, and Soil Pollution* 70: 325-340.

Cairns, M.A. and R.A. Meganck. 1994. Carbon sequestration, biological diversity, and sustainable development: integrated forest management. *Environmental Management* 18(1): 13-22.

CTFS. 1997. Native species carbon sequestration study. Center for Tropical Forest Science. Summer 1997 issue.

Cutright, N.J. 1996. Joint Implementation: Biodiversity and Greenhouse Gas Offsets. *Environmental Management* 20(6): 913-918.

EDF. 1998. Kyoto Climate Agreement Is a Critical First Step. Environmental Defense Fund Electronic Newsletter, April 1998.

Endara, M. Administrator General, National Environmental Authority, Panama City, Panama. Written communication, received August 7, 1998.

Figueres, C., A. Hambleton, and S. Petricone. 1996. Joint Implementation: What? Who? Why? Center for Sustainable Development in the Americas. Washington, D.C.

Goodman, A. 1998. Carbon trading up and running. *Tomorrow*, (May/June): 28.

Greenquist, E.A. 1996. Panama at a new watershed; Panama Canal maintenance and the environment. *Americas* 48(4): 14.

Guardia, T. Panamanian Ministry of Foreign Affairs, Panama City, Panama. Written communication, received August 7, 1998.

Hall, C.A.S. and J. Uhlig. 1991. *Refining estimates of carbon released from tropical land-use change. Canadian Journal of Forest Research* 21: 118-131.

Hanily, G. Project Director, Fundacion Natura, Panama City, Panama. Presentation to Yale students on March 16, 1998, Gamboa, Panama.

Heckadon, S. 1993. The Impact of Development on the Canal Environment. *Journal of Inter-American Studies and World Affairs* 35(3): 129-149.

Houghton, R., V. Dale, A. Grainger, A. Lugo, and S. Brown. 1991. Consequences of Alternative Land Uses for Climate Change. Draft manuscript for the National Research Council Report for the Board of Agriculture's Project on Sustainable Agriculture.

Hubbell, S.P. and R.B. Foster. 1992. Short-term dynamics of a neotropical forest: why ecological research matters to tropical conservation and management. *Oikos* 63: 48-61.

INRENARE. 1994. Brochure on Soberania National Park. Institute of Renewable Natural Resources, Panama City, Panama.

INRENARE. 1990. National Plan for Forestry Action. Institute of Renewable Natural Resources, Panama City, Panama.

Intergovernmental Panel on Climate Change (IPCC). 1996. Climate Change 1995– The Science of Climate Change. Cambridge University Press, England.

LeBlanc, A. 1997. Domestic Greenhouse Gas Trading in the United States. *Global Greenhouse Emissions Trader,* Issue 3.

Programme for Belize (PfB), The Nature Conservancy, and Wisconsin Electric Power Company. 1994. The Rio Bravo Conservation and Management Area, Belize. Carbon sequestration pilot project. Submitted for consideration under the USIJI. The Nature Conservancy, Arlington, VA.

Putz, F.E. and M.A. Pinard. 1993. Reduced-impact logging as a carbon offset method. *Conservation Biology* 7(4): 755-757.

Putz, F.E. and M.A. Pinard. 1996. Retaining forest biomass by reducing logging damage. *Biotropica* 28(3): 278-295.

Smith, W.H. 1993. Forests, Chapter 6. In: Preparing for an uncertain climate. U.S. Congress, Office of Technology Assessment, Vol. II, OTA-0-567, Washington, DC, U.S. Government Printing Office, pp. 299-320.

Tangley, L. 1998. Rain forests for profit: businesses sell nuts, tourism and carbon storage. *U.S. News and World Report.* April 20, 1998.

Trexler, M.C. 1991. Minding the Carbon Store: Weighing U.S. Strategies to Slow Global Warming. World Resources Institute. Washington, DC.

USIJI. 1997. USIJI Uniform Reporting Document: Activities Implemented Jointly under the Pilot Phase. United States Initiative on Joint Implementation, Washington, DC.

USIJI. 1996. Brochure titled "US Initiative on Joint Implementation B Reducing Greenhouse Gas Emissions Through International Partnerships." United States Initiative on Joint Implementation, Washington, DC.

Zollinger, P. and R.C. Dower. 1996. Private Financing for Global Environmental Initiatives. World Resources Institute, Issues and Ideas. Washington, DC.

Rehabilitation
of Former US Military Lands
Bordering the Panama Canal

Claire M. Corcoran

SUMMARY. The Panamanian government is currently negotiating with the United States to determine the extent of the cleanup of US military bases along the Panama Canal. The withdrawal of the US will be completed December 31, 1999. At present, there is no mechanism to ensure that the US will fund, assist with, or assume liability for environmental hazards left on former US Department of Defense lands. Known contamination consists of unexploded munitions on former firing ranges used by the US Army, Navy, and Air Force. Most of these areas are forested with forests of varying age, type, and structure. Complete cleanup could involve complete deforestation of the former bases and firing ranges. The Nature Conservancy has evaluated the ecology of the lands using their "Rapid Ecological Assessment" protocol. The US plans for the lands to become protected areas with warning signs and jersey barriers to protect the public. Panama wants assurance that future cleanup and potential legal damages will be paid for by the US. This paper attempts to summarize the current situation and recommends a potential strategy to both conserve the most biologically diverse forest while maximizing the level of rehabilitation of the areas. It is based on observations and interviews made during a ten-day trip to the Panama Canal Watershed, as well as on news articles, fact sheets from a peace

---

Claire M. Corcoran received a Master of Forestry degree from the Yale School of Forestry and Environmental Studies, New Haven, CT 06511. She is Consultant, Massachusetts Community Forestry Council, Boston, MA 02118.

[Haworth co-indexing entry note]: "Rehabilitation of Former US Military Lands Bordering the Panama Canal." Corcoran, Claire M. Co-published simultaneously in *Journal of Sustainable Forestry* (Food Products Press, an imprint of The Haworth Press, Inc.) Vol. 8, No. 3/4, 1999, pp. 67-79; and: *Protecting Watershed Areas: Case of the Panama Canal* (ed: Mark S. Ashton, Jennifer L. O'Hara, and Robert D. Hauff) Food Products Press, an imprint of The Haworth Press, Inc., 1999, pp. 67-79. Single or multiple copies of this article are available for a fee from The Haworth Document Delivery Service [1-800-342-9678, 9:00 a.m. - 5:00 p.m. (EST). E-mail address: getinfo@haworthpressinc.com].

67

advocacy organization, and primary sources such as correspondence between the US and Panama and minutes from meetings. *[Article copies available for a fee from The Haworth Document Delivery Service: 1-800-342-9678. E-mail address: getinfo@haworthpressinc.com <Website: http://www.haworthpressinc.com>]*

KEYWORDS. Protected areas, military clean-up, unexploded ordnances, Panama Canal

INTRODUCTION

The United States has maintained a military presence in the Republic of Panama since the completion of the Canal in 1914. In accordance with the 1977 Panama Canal Treaty, the US will transfer the 10 mile wide, 648 square mile Inter-Oceanic Region, formerly known as the "Canal Zone," to the Panamanian government at noon on December 31, 1999 (The Panama Canal Treaty 1977). Over the years, operations by the US Army, Navy, and Air Force have damaged the Inter-Oceanic Region environment extensively and created serious health hazards on US military bases (Kovaleski 1998; Wagner 1997). In preparation for the final withdrawal of the US Department of Defense (DOD) from its lands along the west side of the Panama Canal, the Panamanian government and the Panamanian environmental community are negotiating with the US for a clearly defined, long-term policy on the cleanup of the military bases. Cleanup efforts began this year, but because of the seasonal rains, the level of accomplishment before the transfer will be minimal (Kovaleski 1998).

These bases form an almost continuous corridor of forest cover from the Atlantic to the Pacific. The annual level of rainfall in Panama varies along an east-west gradient. The Atlantic coast averages 130 inches of rain per year, whereas the Pacific receives 70 inches of rain per year. This cline results in a high diversity of forest types along the Canal (Graham 1973).

Because much of the bases' contamination is found within the soil in the form of unexploded munitions (referred to henceforth as unexploded ordnances, or UXOs), a thorough cleanup of these areas would involve complete deforestation to gain access to the soil (Hartshorn, pers. comm.). In addition to their great biodiversity, these forests stabilize the soils on steep slopes, preventing erosion and protecting the Canal, including its widest point, Gatun Lake, from siltation (Strieker 1997). The optimal level of cleanup and the establishment of an infrastructure to ensure the health and safety of the Panamanian people is intrinsically uncertain. The combination of the debate over definitions of Treaty language and the lack of information on the extent of the contamination have slowed negotiations considerably.

THE PANAMA CANAL TREATY

The Panama Canal Treaty, signed in 1977 by Carter and Torrijos, which set out the terms of the transfer, contains explicit environmental language mandating full cooperation by both governments in the pursuit of environmental protection. Article VI of the Treaty, entitled "Protection of the Environment," sets up the Joint Commission on Environment (JCE), a bilateral advisory group of citizens that makes non-binding recommendations to each government. Members of the JCE include Mirei Endara, Executive Director of the Institute of Renewable Natural Resources (INRENARE), Gary Hartshorn, the Executive Director of the Organization for Tropical Studies and a tropical ecologist at Duke University, Philip Pillsbury of the US State Department, Dr. Eloy Gibbs, the Panamanian chair of the commission, and Professor Mireya Correa of Panama (JCE 1998a).

Section Three of Article VI of the Panama Canal Treaty commits both governments to providing full and complete information on environmental issues, with ample time to study and act preventively against environmental damage. The Treaty also contains a single phrase on which the current negotiations on environmental protection focus: "The US . . . shall be obligated to take all measures to ensure, insofar as may be practicable, that every hazard to human life, health, and safety is removed . . . " (Panama Canal Treaty 1977). The US is claiming that the remoteness of some areas, the steepness of much of the terrain, the enormous expense of the cleanup, and the danger imposed to personnel by the nature of the contamination make extensive cleanup not practicable (Kovaleski 1998). The DOD also wants to protect the biodiversity found in these forests (US SouthCom Transfer Plan 1997). The Panamanian government, its environmental authority INRENARE, and its agency overseeing the transfer, the Inter-Oceanic Regional Authority (ARI) feel that if the US does not clean these lands, it will be violating its legal and moral obligations under the Treaty (Kovaleski 1998; Lindsay-Poland 1998a).

Unexploded Ordnance

The most widespread environmental hazard on these lands is also the one that would limit their future use the most: thousands of unexploded munitions. UXOs from most US military actions this century litter the Empire, Balboa West, and Piña Ranges. Testing of munitions ability to withstand the tropical heat, wetness, and mold began after the US experience in the Pacific in World War II (The Economist 1998). According to Panama's Foreign Ministry, 21 Panamanian civilians have been killed and many more injured by explosions on or near the ranges of UXOs (Kovaleski 1998). Last year, an official from ARI accidentally detonated three UXOs, though was not injured, while inspecting land near Empire Range (The Economist 1998). Four years ago, a Navy SEAL was seriously injured when a discarded shell exploded under his feet (Kovaleski 1998). In addition to the immediate threat of detonation, INRENARE is concerned about potential health risks from toxic pollutants which may result from UXOs or from other chemical agents that may contaminate the groundwater in these areas (Mirei Endara, pers. comm.).

The Panamanian government has made a formal request to see all documentation on the sites' historical uses (The Economist 1998), yet so far none has been provided (Mirei Endara, pers. comm.). At the February 1998 meeting of the JCE, Rodrigo Noriega, Panama's advisor on environmental affairs to the Foreign Minister, made a presentation on behalf of the JCE's Ad-Hoc Technical Environmental Council, a committee of representatives from the Foreign Ministry's Treaty Affairs Department, INRENARE, and ARI. Noriega emphasized the need for more detailed information on both the bases covered by the Carter-Torrijos Treaty and on bases that reverted earlier. These locales includes San Jose Island in the Pearl Archipelago, which was extensively bombed and where nerve gas was tested, and a small area north of Gamboa which remains mysteriously unforested (JCE 1998a). Noriega proposed that there should be more exchange of information, earlier submission of more complete Installation Condition Reports (ICRs), financing for documentary investigation of the sites, and most significantly, carry-over of US environmental responsibility for the bases after 1999 (JCE 1998a).

Biodiversity

In 1995, The Nature Conservancy did a "Rapid Ecological Assessment" of the military base forests. This exercise collected valuable baseline biological information on much of the area, which had never been scientifically catalogued. Overall, the team of scientists found a wide diversity of forest types across the isthmus, including local areas of extremely high biodiversity. After intensive literature reviews and field surveys of the area's floral and faunal ecology, they ranked each base on a five-part scale of biodiversity significance. Their criteria were biodiversity, condition of the tract, size, number of special elements, degree of fragmentation, and viability of forest patches. They then combined the ranks under each criterion to arrive at a biodiversity rank: outstanding biodiversity significance, very high biodiversity significance, high biodiversity significance, moderate biodiversity significance, or general biodiversity interest (The Nature Conservancy 1995).

The base richest in biodiversity, Fort Sherman, located on 16,774 acres on the Pacific Coast and ranked "of outstanding biodiversity significance," was never used as a firing range. Although the level of UXOs present here is unknown, it is widely assumed that contamination here is minimal (The Nature Conservancy 1995).

The Nature Conservancy ranked the Piña Range "of outstanding biodiversity significance," and found that it is almost entirely forested (The Nature Conservancy 1995). The DOD has stated that 30% of the 6,301 acres was used for "intentional delivery of dud producing ordnance" by the Army (JCE 1998b). The area, located on the northwest shore of Lake Gatun, contains two outstanding patches of tall evergreen forest. It is relatively inaccessible, and as such, there is less human disturbance (poaching, subsistence agriculture by squatters) than there is in other military lands. Its forest cover helps maintain the stability of the banks of the lake and controls erosion and siltation rates (The Nature Conservancy 1995).

The Army also used the 21,938 acre Empire Range for weapons testing, concentrated in roughly 30% of the area. During The Nature Conservancy's assessment, 40% of the land was "restricted access," yet the area still ranked "of very high biodiversity significance" according to their criteria. Aerial photographs were used to determine that the area is covered by three vegetation types, which form very large, mosaic-like tracts of continuous forest cover. The best preserved continuous mixed forest occurs in the Empire Range, and joins the mixed forest of the adjacent Balboa West Range (Figure 1) (The Nature Conservancy 1995). The Empire Range is located within the Canal Watershed, and its forests play an important role in preventing erosion and siltation, and moderating the supply of fresh water (1.6 million gallons per day) that the Canal relies upon for its operation (Greenquist 1996). The *campesinos*, or rural people, who live in or near the Empire Range have been using the land for forest product extraction, hunting, and some subsistence agriculture (The Nature Conservancy 1995).

The situation is similar in the Balboa West Range, which was an Air Force firing range and training center for the Navy SEALS for many years. The land is used by some local *campesinos*. The DOD claims that 34% of the 8,818 acres was used as firing ranges. The Nature Conservancy was denied access to 70% of the land, and so based much of its appraisal on aerial photographs. It found four vegetation types present, and ranked the area as "of very high biodiversity significance," primarily due to the undisturbed nature of much of the forest (The Nature Conservancy 1995).

Chemical Weapons

It is possible that chemical weapons testing may have taken place in Panama. Reports from a variety of sources indicate that some testing took place, but documentation and records on specific tests are lacking (Lindsay-Poland 1998b; The Economist 1998; Kovaleski 1998). Panama is moving toward ratification of the Chemical Weapons Treaty, which would require the US to remove any and all abandoned chemical weapons in Panama, not just those on Canal Treaty lands (Lindsay-Poland 1998a). San Jose Island was

FIGURE 1. Location Map for U.S. Department of Defense Lands in Panama, 1992.

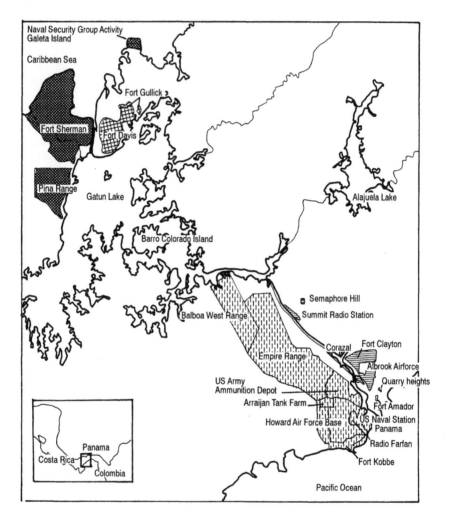

used from late 1944 into the early 1950s to test not only conventional weapons, but also mustard gas and other chemical weapons. Another non-Treaty area is Chivo-Chivo, an area restored to Panama in the 1970s. A former consultant for the DOD, Rick Stauber, claims to have seen documents that show a "function test"–a firing test–of a nerve gas landmine there. Other sources say that almost every type of non-nuclear weapon used since the 1940s, including napalm and Agent Orange, were at some point tested in Panama (The Economist 1998).

A laboratory facility known as the Tropic Test Center (TTC), located on Corozal base, admitted to conducting some "storage testing" of nerve gas from 1964 to 1968, in which the projectiles were stored to evaluate the effect of the tropical climate on them. After a US-based peace group called the Fellowship of Reconciliation (FOR) obtained a Pentagon study of firing ranges in Panama under the Freedom of Information Act, the Panamanian government learned of the use of depleted uranium antitank projectiles (DU) on the Empire Range (Lindsay-Poland 1998b).

The US Southern Command (SouthCom), the DOD branch that administered the bases, has claimed that the TTC conducted only "storage tests" of DU projectiles in 1993-1994. Rick Stauber (Lindsay-Poland 1998b) claims to have obtained TTC records that showed that the DU and nerve gas weapons were fired on the Empire Range. His documentation shows that depleted uranium weapons were fired in the early 1980s (Lindsay-Poland 1998b). The Panamanian Foreign Minister Ricardo Alberto Arias stated that Panama never authorized chemical testing on its territory, and that his government was also investigating the issue. Unauthorized chemical testing would represent a transgression of the treaty terms and of other international accords. American officials have said that few records were kept of activities at the bases, and that many have been lost since (Lindsay-Poland 1998b).

Firing Ranges as Nature Preserves

In October of 1997, SouthCom submitted a plan for the cleanup and transfer of the bases. The US proposes to turn most of the contaminated land into protected areas, suitable for limited recreational use (US SouthCom Transfer Plan 1997). This idea uses as a model many base closures in the United States, and the management of many World War II battlefields in France which are now war monuments. US examples include Fort Ord, California; Fort Meade, Maryland; and Fort McClellan, Alabama. Early in 1988, the US Ambassador to Panama sponsored a tour of these US sites for a group of Panamanian government officials, scientists, and environmentalists to see how this model is implemented in the US (JCE 1998a).

The plan excludes cleanup of any chemical weapons, and makes no commitment to any cleanup after the US withdrawal. It calls for the surface cleanup of less than 1% of contaminated areas. No forested lands or steep terrain would be cleared as these are deemed inaccessible. In grassland areas of "none, suspected, or very low UXO densities," the plan calls for the "development of an action plan" to clear the surface or partially buried UXOs. In areas of low through very high density of UXOs, jersey barriers would be placed along roads to restrict vehicular access, and warning signs would be posted along roads and foot trails through the forest. Warning signs would also adorn all accessible areas and would include a floating buoy sign at an inlet on Balboa West (US SouthCom Transfer Plan 1997).

No below-ground cleanup is planned because of high expense, loss of flora and fauna, and hazard to personnel. The DOD also cites the fact that the technology to differentiate between buried scrap metal and UXOs does not exist. According to the plan, 307 signs would be installed at Empire, 300 at Balboa, and 183 at Piña. Col. Michael DeBow, deputy chief of staff of US SouthCom, said that the US military is committed to an educational program to accompany the signs (US SouthCom Transfer Plan 1997).

Cultural Context

There may be some inherent difficulty in implementing the traditional American/European model of the firing range or battle ground as nature preserve in rural Panama. For example, there is a legal and cultural practice of land acquisition in Panama that does not occur in the US. Citizens can obtain permanent legal tenure of land through proven long-term use; what is known as "squatting" in the US is common and legal in Panama (Economist Intelligence Unit Profile 1995). The personal safety of squatters who settle in the proposed park system will be compromised by the presence of UXOs. Also, although the literacy level is over 90% for people over 15 years old, literacy is often lower in rural areas with higher poverty levels (Economist Intelligence Unit Profile 1995). In Panama, the poverty level in rural areas is 63%, with 42% described as "extremely poor" (The Economist, 1998).

ARI has rejected the initial SouthCom transfer and cleanup plan, citing as reasons the small size of the area to be cleaned–less than .22% of contaminated areas–and the vast areas that would be left permanently useless (Barletta 1997). The SouthCom plan called for cleaning a total of 548 acres of the 6,000 acres of the most heavily contaminated lands (US SouthCom Transfer Plan 1997).

The Cleanup

The JCE, as well as ARI, is concerned about ongoing liability issues relating to the cleanup. An estimated 20,000 people currently inhabit small towns surrounding the ranges, a number expected to grow to 100,000 over the next 25 years. Despite existing signs warning, "EXPLOSIVES KILL ANYBODY, DO NOT TRESPASS," residents cross the range borders to fish, collect forest products and scrap metal, and plant crops (Kovaleski 1998).

Gary Hartshorn, an American tropical ecologist who is the co-commissioner of the JCE, said he is working hard to ensure that there is a mechanism for long-term accountability, legally and financially, for any loss of life incurred by Panamanian citizens as a result of US military actions. He also added that in a recent JCE helicopter flight over the ranges, visual inspection discovered little new settlement or agriculture within their borders (Hartshorn, pers. comm.).

Although previous base cleanups typically take years to accomplish, the effort in Panama began two years before the transfer, leaving two dry seasons for the US military to complete its environmental work. The US Ambassador to Panama, William J. Hughes, who describes this issue as "probably one of the stickiest issues we are dealing with here," has stated that he believes the US can comply with its environmental obligations before December 31, 1999. However, he admits that ". . . the parameters of what those responsibilities are vague" (Kovaleski 1998).

Thus far in the 1998 dry season, the military has cleaned almost 90 acres, destroying 250 explosives and removing an estimate 80,000 pounds of scrap metal. It has also been conducting ICRs, to gather more information on the extent of the contamination. The US has stated, "the cleanup got off to a slow start. It was not on the radar screen" (Kovaleski 1998). The Panama Canal Commission (PCC), the US agency that oversees Canal operations, has contracted the Army Corps of Engineers to conduct clearance operations in areas in two ranges, where soil from construction of the third set of locks is to be deposited (Lindsay-Poland 1998a). Balboa West was closed in January so that cleanup could proceed. In addition, ARI has contracted with a Canadian tribe, the Tsuu T'inas, to clean over 750 heavily contaminated acres in Empire Range. The tribe is experienced in this kind of cleanup, having gained their expertise cleaning their ancestral lands which had been used as a Canadian firing range (Lindsay-Poland 1998a; Kovaleski 1998).

One proposal would convert the Corozal base, along with parts of the Empire and Piña Ranges and Howard Air Force Base, into a multinational anti-drug trafficking center. The scheme, which would involve an American military presence, is in the negotiating phase (Lindsay-Poland and Isacson 1998). However, according to JCE co-commissioner, Gary Hartshorn, the political support in Panama is severely lacking for this type of venture (Hartshorn, pers. comm.). If it were approved, though, it could necessitate a more extensive cleanup of those areas.

ETHICAL AND LEGAL IMPLICATIONS

The Fellowship of Reconciliation believes that the US military is intentionally avoiding its responsibilities. John Lindsay Poland, FOR's director of Latin American programs, said, "In Panama, the Pentagon does not want to set a precedent for the cleanup of overseas bases, despite its obligations" (Kovaleski 1998). The group, along with the Coordinadora Popular de Derechos Humanos de Panama (Panamanian Popular Coordination of Human Rights), solicited an *amicus* brief from the Sierra Club Legal Defense Fund (SCLDF, now the Earthjustice Legal Defense Fund) on the legal issues involved in the cleanup.

International Law

The brief, prepared by SCLDF and Heller, Ehrman, White and McAuliffe, summarizes the principles of international law which apply to this situation. The brief states the obligations of the US under international law. According to the ethical principles of comity, moral responsibility, and non-discrimination, the US ought to apply the same standards to base closings in other countries as it does domestically; the environmental laws which apply in domestic base closures, such as National Environmental Protection Act, the Resource Conservation and Recovery Act, and Comprehensive Environmental Response Compensation and Liability Act should be followed in US base closures abroad (Wagner 1997).

The SCLDF legal team also concludes that by attempting to interpret the phrase "to the extent practicable" in a narrowly defined sense without the agreement of the Republic of Panama represents an attempt to redefine an unambiguous Treaty term without the consent of the other Treaty parties, in violation of specific international legal rule. The US has affirmative obligations to protect the environment, which derive from treaties, customary international law, and general principles of international law. The brief concludes that the US has disregarded its obligation under such law to not harm the environment of another state, as well as obligations regarding particular types of environmental problems–hazardous wastes, chemical weapons, and activities that endanger biological diversity–and obligations to human rights principles, most importantly the right to self determination (including sovereignty over natural resources). The brief concludes: "the current situation in Panama reflects an unacceptable disdain for the rule of law and a flagrant disregard for the rights of the people of Panama" (Wagner 1997).

CONCLUSION

Panama's environmental officials are trying to secure commitments for long-term solutions to the contamination. ARI's *de Grimaldo* said recently,

"The environmental responsibility of the United States should not end in 1999; the two governments should agree on a mechanism to extend that responsibility" (Kovaleski 1998). The JCE is attempting to develop and flesh out a strategy to present to the two governments that will be acceptable to both. To date, it has identified four key areas: the gathering of scientific, technical, and historical information; the monitoring of upcoming projects and the cleanup of the bases; a mechanism for follow-up to promote protection of the environment after the Treaty concludes in 1999; and the development of the financing needed to make such protection possible. The committee is exploring the possibility of creating a trust fund that would cover future cleanup operations and liability issues (JCE 1998b).

Recommendations

The area of Fort Sherman is the most worthy of protection, as it contains 2,548 acres of *cativo*, or flooded forest, one of Panama's most threatened forest types. This base also earned The Nature Conservancy's sole ranking of outstanding biodiversity significance, because of the *cativo*, a 282 acre patch of the extremely rare tropical deciduous forest type, and the best examples of "tall forest" (relatively undisturbed forest over 200 years old) in all the bases (The Nature Conservancy 1995). It is also one of the few large areas that is presumed safe to develop for industrial, agricultural, or residential use. The high commercial value of the *cativo* forest, combined with the scarcity of safe land on former US lands, increases the likelihood that the land will be cleared and converted to another use. The US has the opportunity to facilitate Fort Sherman's protection by cleaning other areas to acceptable levels for industrial, if not residential, use.

The FOR has proposed an alternative method of cleanup that would mitigate the negative impacts of the deforestation necessary to clean the soil of UXOs. It suggests cleaning the forested lands in a checkerboard pattern, assuring the proximity of vegetation that would re-colonize the cleared area, and minimizing the erosion resulting from massive soil disturbance on steep slopes. The US should explore the feasibility of this option thoroughly, as it may provide a balanced solution between biodiversity protection and cleanup (Lindsay-Poland 1998c).

The US also should commit formally to a continuing effort to develop technology to rehabilitate UXO-contaminated lands, and to transfer such technology to the Panamanian government after US withdrawal. Without a firm, clear, and financially backed statement of the US intention to facilitate complete cleanup of former US military bases, the integrity of its international relations with all of Central America will suffer.

At this point there is no concrete commitment by the US to any course of action. The US could avoid a potential disaster, ecologically and politically, if it were to commit to funding long-term cleanup and assuming liability for future damages. As a developing tropical nation, Panama must have complete sovereignty over its resource use and future development. The continuing commitment of the US to fulfill the spirit and intent of the Treaty will be necessary for the security of Panama's future.

REFERENCES

Barletta, N. Unpublished. Letter to Minister of Foreign Relations Ricardo Alberto Arias. Re: Plan de Transferencia de los Polligonos de Tiro en el Area del Canal. 5 March 1997.

Economist Intelligence Unit Profile. 1995. Country Profile 1994-1995: Panama. Economist Intelligence Unit Ltd. UK.

The Economist. Panama: Clearing Up After Uncle Sam. 28 February 1998.

Endara, M. Director, IRENARE, Panama City, Panama. Discussion on the Panama Canal Watershed, March, 1998, Gamboa, Panama.

Graham, A. 1973. Vegetation and vegetational history of northern Latin America. Elsevier Scientific Publishing Company, New York.

Greenquist, E.A. 1996. Panama at a new watershed: Panama Canal maintenance and the environment. *Americas* 48(4): 14.

Hartshorn, G. Executive Director, Organization for Tropical Studies. Discussion on US military lands in Panama. March, 1998. Panama City, Panama.

Joint Commission on Environment a. Agenda, XXVI Meeting, 18 March 1998.

Joint Commission on Environment b. Minutes of the XXV Meeting, February 19-21, 1998.

Kovaleski, S.F. A Dangerous American Legacy: Acres of US Military Land in Panama Are Littered with Unexploded Munitions. *Washington Post* 2 April 1998: A27.

Lindsay-Poland, J. a. "Advances in US Commitment to Cleanup . . . and Some Continuing Problems." Fellowship on Reconciliation 21, December, 1997. On-line. Internet. 4 May 1998. Available http://www.nonviolence.org/for/panama

Lindsay-Poland, J. b. "Depleted Uranium and Nerve Gas: 'A Hornets Nest.' " Fellowship on Reconciliation 20, Summer, 1997. Online. Internet. 4 May 1998. Available http://www.nonviolence.org/for/panama

Lindsay-Poland, J. c. "Pentagon Study of Ranges in Panama Reveals Explosive Problems." Fellowship on Reconciliation 19, Spring, 1997. Online. Internet. 4 May 1998. Available http://www.nonviolence.org/for/panama

Lindsay-Poland, J. and A. Isacson. "Military Negotiations Stalled Despite 'Agreement in Principle.' " Fellowship on Reconciliation 22, March, 1998. Online. Internet. 4 May 1998. Available http://www. nonviolence.org/for/panama

The Panama Canal Treaty, Article VI, sections 1-3. 1977.

The Nature Conservancy. Rapid Ecological Assessment of US Military Lands in Panama, Phases III and IV. 1995.

Strieker, G. Forests Along Panama Canal Face Uncertain Future. November 18, 1997. CNN, Atlanta, GA. Online. Internet. Available: http://forests.lic.wisc.edu/gopher/centamer/panforest

US SouthCom Transfer Plan, Balboa West, Empire, and Piña Ranges. October 1997, US Department of Defense, Panama.

Wagner, J.M. Environmental Injustice on US Bases in Panama: International Law and the Right to Land Free from Contamination and Explosives. Sierra Club Legal Defense Fund and Heller, Ehrman, White and McAuliffe, on behalf of Fellowship on Reconciliation and Coordinadora Popular de Derechos Humanos de Panama. July 1997. San Francisco, CA.

Sedimentation
in the Panama Canal Watershed

Maya Loewenberg

SUMMARY. Research on sedimentation rates in water supply reservoirs of the Panama Canal Watershed is reviewed. Factors that influence sedimentation rates are discussed, including topography, climate, precipitation intensity, land use (e.g., conversion from forest to agriculture), and certain rare events (e.g., landslides). Land use is the most significant factor affecting regional and global variations in sediment yield. This discussion focuses on Alhajuela Reservoir, which supplies water to the Panama Canal. Insufficient data are available for accurate predictions of storage loss rates by sedimentation in the Alhajuela Reservoir, in large part because of the evolving land use in the region. Moreover, experience shows that inference of storage loss rates from supposedly similar reservoirs elsewhere is unreliable due to the complex and poorly understood multivariate dependence of the problem. Two methods were used to estimate sediment yield, as correlated with reservoir size and watershed area. Predictions obtained by these methods diverge, so their general reliability is therefore questionable. Ongoing studies should facilitate reliable predictions of sedimentation rates. *[Article copies available for a fee from The Haworth Document Delivery Service: 1-800-342-9678. E-mail address: getinfo@haworthpressinc.com <Website: http://www.haworthpressinc.com>]*

KEYWORDS. Panama Canal, sediment yield, watershed management, erosion

Maya Loewenberg recently completed a Master of Environmental Studies degree at the Yale University School of Forestry and Environmental Studies, New Haven, CT 06511.

[Haworth co-indexing entry note]: "Sedimentation in the Panama Canal Watershed." Loewenberg, Maya. Co-published simultaneously in *Journal of Sustainable Forestry* (Food Products Press, an imprint of The Haworth Press, Inc.) Vol. 8, No. 3/4, 1999, pp. 81-91; and: *Protecting Watershed Areas: Case of the Panama Canal* (ed: Mark S. Ashton, Jennifer L. O'Hara, and Robert D. Hauff) Food Products Press, an imprint of The Haworth Press, Inc., 1999, pp. 81-91. Single or multiple copies of this article are available for a fee from The Haworth Document Delivery Service [1-800-342-9678, 9:00 a.m. - 5:00 p.m. (EST). E-mail address: getinfo@haworthpressinc.com].

BACKGROUND

Deforestation and urbanization within the Panama Canal Basin may imperil the future operation of the Panama Canal, as well as the quality and availability of water supplied to Panama City. The Smithsonian Tropical Research Institute (STRI) was funded in 1996 by the US Agency for International Development (USAID) to assist Panama's Institute for the Management of Renewable Natural Resources (INRENARE) in establishing an environmental monitoring program with a focus on monitoring the effects from the land uses in the Canal Watershed. Data from the monitoring program are intended to enable INRENARE to rationally select and implement proven land-use policies.

The phenomenon of sediment discharge regimes to reservoirs and river systems has been studied in the temperate climates for decades, but little data exist for humid tropical regions (Hamilton and King 1983). Watershed experiments are generally long-term, expensive, and require a high degree of technical competence. Thus, most research has focused on wealthy, temperate, industrial countries. Furthermore, direct transfer of research to an entirely different climatic zone is fraught with uncertainty. It must be recognized that both the nature of the driving forces (e.g., precipitation intensity or temperature) and the response of a system (e.g., sediment yield or evapotranspiration loss) may be quite different in the tropics (Hamilton and King 1983).

The intent of this paper is to review literature on reservoir sedimentation in tropical and temperate climates. The discussion is focused on the storage depletion rate of the Alhajuela Reservoir resulting from sedimentation.

IMPORTANCE OF ALHAJUELA WATERSHED

Water Demand

Large amounts of fresh water–about 62% of the total runoff–are discharged through the locks to the Atlantic and Pacific oceans in the daily operation of the Panama Canal. This water is supplied by runoff from the 3339 km^2 watershed of the Chagres River, also known as the Canal Watershed. The Chagres River yields about 2.8 billion gallons of fresh water daily. Each ship passing through the canal requires about 52 million gallons of water. The next largest users of water from the Chagres are hydroelectric power dams, which require 32% of its daily yield. The remaining 6% of the daily yield of fresh water is put towards municipal use (Heckadon 1993). It is estimated that by the year 2000, almost 4.0 billion gallons of fresh water will have to be extracted daily from the Chagres River basin to satisfy increased water demand (Heckadon 1993).

Description of Alhajuela Watershed

The Canal Watershed is divided by the Madden Dam into the upper and the lower basins: Alhajuela and Gatun, respectively. Water storage for Canal operation is provided by two reservoirs, Lake Gatun and Lake Alhajuela. The upper Alhajuela Watershed, although only one-third of the drainage area, provides almost half of the Canal's water because of the area's mountainous topography and high rainfall. Therefore, the Alhajuela Reservoir is more important than the lower Watershed as a water source for the Panama Canal. Moreover, from a sedimentology point of view, the Alhajuela Lake Watershed is more critical to the Canal than the Gatun Lake Watershed. The Alhajuela Watershed is considered more susceptible to erosion due to steep slopes, storm trajectory, rainfall intensities, and soil type (Alvarado 1985). Lake Alhajuela serves also as the water supply for Panama City (700,000 people).

The Alhajuela Watershed (975 km^2) has a highly developed drainage pattern with narrow valleys and steep, irregular slopes (see Figure 1). Land slopes exceed 45% in 92.4% of the Watershed area, which makes it unsuitable for grazing or farming (Larson 1984). The most representative soil type in the upper Watershed is red clay under a thin layer of humus. This type of

FIGURE 1. The steep irregular slopes of the Chagres National Park located in the Alhajuela Watershed. Photo credit: Jennifer L. O'Hara.

soil is very likely to erode under intense rainfall conditions and steep slopes, even under forest cover (Foster, 1973). The most important part of the Watershed is the Chagres forest reserve, which encompasses approximately 750 km^2 of forest (based on 1983 estimate). Currently, work is underway to obtain an accurate estimate of forest cover and land uses within the Watershed.

The mean annual rainfall in the Alhajuela Watershed between the years 1982 and 1996 was 98 inches (Alvarado 1998). However, prior to 1982, an annual 113 inch average in precipitation was observed (Alvarado 1985). The cause of this 13% decline is unknown.

SEDIMENTATION PROBLEM IN THE CANAL WATERSHED

Understanding sediment yield to the reservoirs is essential for estimating reservoir life span far in advance. Hydrologists and engineers of the Panama Canal have long been aware that changes in land uses throughout the drainage basin will induce changes in the content of fluvial sedimentation of the rivers and lakes in the Panama Canal Watershed. When the Alhajuela Reservoir was created in 1934, it was estimated that it would take 3900 years to deplete the storage capacity of the reservoir by 26% (Tutzauer 1990). That projection was based on conditions in 1930s, but changes that have occurred render that projection invalid.

In 1975, it was estimated that 5% of the active storage (the portion of the reservoir that is above the outlet) of Lake Alhajuela (612 $\times$ 10^6 m^3) had been lost by sedimentation (Larson 1979). Meanwhile, visual observations indicated that sedimentation of the reservoir was increasing rapidly due to colonization and deforestation. In 1978, USAID initiated a project that investigated sedimentation rates during two periods: pre-development (1934-73) and post-development (1973-78). It was found that the land clearing, in only small portion of the watershed, had caused a 177% increase in sedimentation (Larson 1984). If erosion and sedimentation were to continue at the 1973-78 rate, the loss of storage capacity in Lake Alhajuela would be 22.6% by the year 2000 and 46.4% by the 2020 (Larson 1979).

After alarming publications by Larson (1979) and by Wadsworth (1978) who forecasted a 40% storage capacity loss of the lake due to siltation, the Panama Canal Commission in the 1980s started a suspended sediment investigation program of major affluent rivers to the Panama Canal. The study by Alvarado (1985) showed that rates of storage capacity loss averaged 0.26% per year. By the year 2035, 100 years after construction, the loss would be as high as 26%. However, if deforestation continues at the rate estimated in 1983 (8 ha/yr) and reaches the Chagres Reserve, sedimentation inflow to the reservoir will increase (Alvarado 1985).

SOURCES OF SEDIMENTATION

STRI scientists hypothesize that significant siltation of the lakes within the watershed is due to two processes: (1) deposition of mineral sediments caused by the erosion of soils, and (2) deposition of organic sediments caused by biological productivity within the lakes (STRI et al. 1997). Deforestation and agricultural development may promote the first process by accelerating physical erosion over natural levels and may promote the second process through the release of nutrients from logging. However, nutrients that are released to rivers and lakes due to agriculture and urbanization are the most serious contributors of organic sediments.

Sedimentation Due to Landslides

Landslides triggered by heavy rainfalls are a historic problem for the Canal and are particularly serious in the steeper portions of the Watershed. Between 1989 and 1993, the Panama Canal Commission dredged about 11 $\times$ 10^6 m^3 of sediment and rock from the Canal, which is 6.3% of the volume excavated during the construction of the Canal (Greenquist 1996). According to the Commission representatives, debris slides along the steep slopes of the Canal caused most of this sedimentation. Much of the Alhajuela Watershed is similarly prone to landslides, which may contribute significantly to the average sedimentation rate.

Landslides are sensitive to land use. The frequency of landslides in susceptible areas may increase as the result of land use conversion. Thus, landslides may be a further manifestation of increased sedimentation rates resulting from land use conversion.

Land Use and Sediment Yield

Analysis of sediment yields by Dunne (1979) from 61 Kenyan catchments revealed that land use is the most important factor affecting regional and global variations in sediment yield. Dunne (1979) found that sediment yields in undisturbed Kenyan catchments are relatively low throughout a range of climate. Under a single land use, however, it is possible to recognize the effect of topography, runoff and rainfall intensities.

According to Dunne (1979), sediment yield from forested regions is an average of 20 t km^{-2} yr^{-1}. Where rainfall intensities are higher, average sediment yield in forested regions increases to 30 t km^{-2} yr^{-1}. Much greater sediment yield was observed on cultivated and grazed lands. Measured values range from 1,000 to 5,000 t km^{-2} yr^{-1}. These high values of sediment yield are mostly the result of intense tropical rainstorms. Runoff and erosion in agricultural areas are most intense on dirt roads and tracks and contribute approximately up to 35% of the total sediment yield (Dunne 1979). Therefore, erosion from rural roads and footpaths may generate a large part of the sediment yield of catchments and may cause an over-estimation of the degradation of cultivated lands.

The conversion of forest to agricultural fields may significantly enhance erosion, according to a study by Larsen. Larsen (1995) attempted to quantify the increase in sediment yield resulting from land use conversion in two pairs of humid-tropical montane watersheds in Puerto Rico. He found as the result of conversion from forest to agriculture, annual suspended sediment yield increased by a factor of five.

Sediment Yield and Drainage Area

A study by Dendy and Bolton (1976) showed that average sediment yields are proportional to the 0.16 power of drainage area. As drainage area increases, average land slopes usually decrease. This observation qualitatively explains why sediment yield increases less than linearly with drainage area. The relationship described by Dendy and Bolton, however, provides an approximation of mean sediment yields for preliminary watershed planning.

Sediment Yield from Lake Alhajuela Sub-Watersheds

Three sediment-laden rivers flow into the Alhajuela Reservoir from Alhajuela sub-watersheds: the Boqueron, Pequeni and Chagres (Figure 2). The sediment yield contributions from these rivers is shown in Table 1. The Boqueron and Pequeni have the highest annual sediment yield rate: 500-1,000 ton km^{-2} during 1981-1994 (STRI et al. 1997). The high sediment yields of these rivers may be the result of more intensive land uses, including pasturelands and cultivated areas, in these sub-watersheds. In contrast, the Chagres River sub-watershed consists mostly of primary forest. In total, the three rivers contribute 90 percent of the total sediment yield in Lake Alhajuela, according to the results listed in Table 1. In addition, there are 15 ungauged streams that carry approximately ten percent of the total waterborne sediment deposited in the reservoir. Most of these streams are not year-round contributors.

SEDIMENTATION IN ALHAJUELA COMPARED TO US RESERVOIRS

According to Alvarado (1998), the sedimentation rate of the Canal reservoirs is not alarming. He cites publications by Dendy and Champion (1973),

FIGURE 2. Sedimentation rate (ton km^2 yr^{-1})

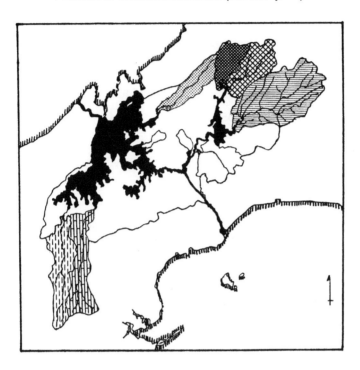

Watershed boundary

☐ No data

▦ 92–Trinidad river sub-watershed

▦ 115–Ciri Grande river sub-watershed

▤ 255–Chagres river sub-watershed

▦ 293–Gatun river sub-watershed

▨ 664–Pequeni river sub-watershed

▨ 870–Boqueron river sub-watershed

Map and data based upon the Panama Canal Watershed Ecological Monitoring project Annual Technical Report (1997) by INRENARE.

TABLE 1. Rivers flowing into Lake Alhajuela, their sub-watershed area, sediment yield, and percent of total yield.

River	Sub-Watershed Area* (km^2)	Sediment Yield[♦] ($t\ km^{-2}\ yr^{-1}$)	Total Yield (%)
Boqueron	91	870	44
Pequeni	135	664	33
Chagres	414	255	13
Ungauged streams	n/a	201	10
Total	640	1990	100

* Information from STRI et al. 1997.
♦ Total waterborne sediment volume from Alvarado (1985) was converted to mass using density 1.6 t/m^3.

and Dendy and Bolton (1976) that indicate the median sedimentation rate in US reservoirs of comparable size to the Alhajuela Reservoir to be five times higher.

Dendy and Champion (1973) compiled sedimentation data for 1,105 reservoirs in the US. The intent of their paper was to provide insight into US reservoir siltation problems. Dendy and Bolton (1976) correlated reservoir sedimentation in the US primarily to two factors: reservoir size (storage capacity) and effective contributing area. The effective contributing area is the area upstream of the reservoir that contributes sediment. The contributing area for Alhajuela Reservoir is the whole upstream Watershed.

Alvarado (1985) and Tutzauer (1990) made their prediction using the first method. Sedimentation in Alhajuela Reservoir is tabulated for the active and inactive portions of the reservoir. The inactive portion of the reservoir is the volume below the depth of the reservoir outlet. According to Alvarado and Tutzauer, capacity loss of the active portion of the Alhajuela Reservoir is one-third the rate of the typical US reservoir during the period 1933 to 1983. However, combined storage loss rate is one half of the typical US reservoir.

An alternative approach is based on effective contributing area. Accordingly, the amount of sediment deposited per year in the reservoir is divided by the watershed area (exclusive of the lake). The derived annual sediment yield in Alhajuela Reservoir is 943 m^3/km^2. According to the data estimated by Dendy and Bolton (1976), US reservoirs with similar contributing areas have annual sedimentation yields of about 250 m^3 per square kilometer of contributing area. Therefore, the sedimentation accumulation in Alhajuela is about four times the rate of reservoirs with similar effective contributing areas in the US. The higher rate of sedimentation is perhaps attributable to the high intensity and erosivity of tropical rain (Stallard, pers. comm.).

Apparently, predictions are very sensitive to the prediction method. The wide range in deposition rates in reservoirs of similar size and drainage area, even within a given land resource area, suggests that average rates would be of little value in predicting the useful life of a particular reservoir.

SOIL CONSERVATION

The issue of soil conservation is closely related to that of land use. Thus, research on soil conservation has direct implications for predicting and controlling storage loss in reservoirs by sedimentation.

A firm commitment to a philosophy that the soil should be conserved is essential to solving erosion and sedimentation problems. Involving local community leaders in soil conservation and raising their awareness of erosion's dangers is imperative. It is important that land users recognize that erosion inevitably reduces soil productivity (Lal 1985).

In an attempt to curb damaging soil erosion in the Canal Watershed, a US development organization called TechnoServe is teaching rural community leaders the basics of environmentally friendly farming. After the training, community leaders hold workshops in their neighborhoods and pass on information to farmers about organic farming, soil conservation, and watershed protection.

The short- and long-term aspects of soil conservation activities need further investigation to provide governments in developing tropical countries with better knowledge for planning and investment and to give farmers a sound basis for participation in soil conservation programs.

CONCLUSION

Research on the sedimentation rate in Alhajuela Reservoir has been reviewed. This reservoir is a very important water supply for the Panama Canal and for Panama City. Several important factors have been identified including topography, climate, precipitation intensity, land use, and certain rare events (e.g., landslides). Land use is the most significant factor. Thus, soil conservation is important. Insufficient data are available for accurate predictions of storage loss rates by sedimentation in the Alhajuela Reservoir because of the evolving land uses in the region. Two methods are popularly used for predicting sedimentation yield in lakes and reservoirs; however, very different results are obtained by the two methods, casting doubt on their reliability.

Literature describing research on reservoirs elsewhere has also been surveyed. However, inference of storage loss rates from supposedly similar reservoirs elsewhere is unreliable due to the complex and poorly understood multivariate dependence of the problem. A regression analysis based on the available data is infeasible.

The problem of erosion and sedimentation is very important in some areas of the world and has caused much concern for the future of water and land resource development in these areas. The literature reviewed in this paper highlights the need for further research on sedimentation in reservoirs.

New research information is becoming available for countries in the humid tropics and will guide natural resources managers in implementing best management strategies. Data on sedimentation in Panama Canal Basin is quite advanced. For example, the main rivers in the Panama Canal Watershed have daily sediment measurements unlike most other rivers in the tropics. Another tropical location that has received much attention is Puerto Rico.

Increased understanding of the many factors determining the amount of erosion in the tropics, on different soil types and under different soil management methods, will help managers to design secure, economically feasible, and readily applicable control systems. The main challenge in implementing control systems is social constraints presented by the local communities who derive their subsistence from producing food crops in areas of high erosion hazard.

REFERENCES

Alvarado, L.A. 1985. Sedimentation in Madden Reservoir: Balboa Heights, Panama: Meteorological and Hydrological Branch, Engineering Division, Panama Canal Commission.

Alvarado, L.A. 1998. The Panama Canal Watershed: Past, Present and Future Panama Canal Commission, Engineering and Industrial Services, Meteorological and Hydrographic Branch.

Dendy, F.E. and G.C. Bolton. 1976. Sediment yield-runoff drainage area relationships in the Untied States. *Soil and Water Conservation* 12:264-267.

Dendy, F.E. and W.A. Champion. 1973. Reservoir sedimentation in the United States. Misc. Pub. 1266. U.S. Department of Agriculture, Washington, DC.

Dunne, T. 1979. Sediment yield and land use in tropical catchments. *Hydrology* 42: 281-300.

Foster, G.R. 1973. Evaluating irregular slopes for soil loss prediction. American Society of Agricultural Engineers.

Hamilton, L.S. and P.N. King. 1983. Tropical Forested Watersheds. Westview Press, Boulder, CO.

Heckadon, S. 1993. The impact of development on the Panama Canal environment. *Journal of Inter-American Studies and World Affairs.* 35(3), 129-149.

Greenquist, E.A. 1996. Panama at a new watershed; Panama Canal maintenance and the environment. *America* 48(4): 14.

Lal, R. 1985 Soil erosion and its relation to productivity in tropical soils. In pp. 237-248. S.EL-Swaify (ed.) Soil erosion and conservation. Soil Conservation Society of America, Ankeny, Iowa.

Larsen, M.C. 1995. Comparison of sediment yields from forested and agriculturally-enveloped montane humid-tropical watersheds, Puerto Rico, International Association of Geomorphologists-Southeast Asia Conference, Singapore, June 18-23.

Larson, C.L. 1984. Controlling erosion and sedimentation in the Panama Canal Watershed. *Water International* 9: 161-164.

Larson, C.L. 1979. Erosion and sediment yield as affected by land and slope in the Panama Canal Watershed. In: Proceedings of III World Congress on Water Resources. pp. 1086-1095.

Stallard, R. Professor of Geology, University of Colorado at Boulder, Boulder, CO. Telephone conversations on sedimentation rates in Panama, April, 1998.

STRI, USAID, INRENARE 1997. Panama Canal Watershed ecological monitoring. *Annual Technical Report.* 70 pp.

Tutzauer, J.R. 1990. Madden Reservoir Sedimentation, 1984-1986: Balboa Heights, Panama, Meteorological and Hydrological Branch, Engineering Division, Panama Canal Commission.

Wadsworth, F.H. 1978. Deforestation–Death to the Panama Canal. In: Proceedings of the US Strategy Conference on Tropical Deforestation, Washington, DC.

Will the Goals Be Met?
An Examination
of ARI's General and Regional Plans
with Respect to Protected Areas
Within the Inter-Oceanic Region, Panama

Sarah N. Whitney

SUMMARY. US government land along the Panama Canal, known as the Inter-Oceanic Region, will revert to Panamanian control on December 31, 1999. The Inter-Oceanic Regional Authority (ARI) was established to allocate land use and encourage development within the region. ARI produced both general and regional plans that were adopted by the government as law in 1997. While Panama has created a model that tries to balance economic development and natural resource protection, the plans lack clarity and specificity about how the natural resource protection efforts will be implemented and evaluated. In particular, the plans should be specific about who will manage the newly protected areas and what the management goals for these areas will be. An evaluation mechanism must be established to determine if the plans are achieving these goals. Because ARI's plans are deficient in these areas, Panama will be significantly hindered or rendered unable to achieve its goal of natural resource protection within the Inter-Oceanic Region. There is still time before the land reversion occurs to remedy

---

Sarah N. Whitney received a Master of Forestry degree from the Yale University School of Forestry and Environmental Studies and is Forester for the Society for the Protection of New Hampshire Forests, Concord, NH.

[Haworth co-indexing entry note]: "Will the Goals Be Met? An Examination of ARI's General and Regional Plans with Respect to Protected Areas Within the Inter-Oceanic Region, Panama." Whitney, Sarah N. Co-published simultaneously in *Journal of Sustainable Forestry* (Food Products Press, an imprint of The Haworth Press, Inc.) Vol. 8, No. 3/4, 1999, pp. 93-105; and: *Protecting Watershed Areas: Case of the Panama Canal* (ed: Mark S. Ashton, Jennifer L. O'Hara, and Robert D. Hauff) Food Products Press, an imprint of The Haworth Press, Inc., 1999, pp. 93-105. Single or multiple copies of this article are available for a fee from The Haworth Document Delivery Service [1-800-342-9678, 9:00 a.m. - 5:00 p.m. (EST). E-mail address: getinfo@haworthpressinc.com].

93

these problems. This can be done by increasing the specificity of the regional plan as it relates to protected areas. At the same time, evaluation procedures need to be established to examine both biophysical changes within the protected areas and the function of agencies involved with natural resource management in the Inter-Oceanic Region. If these measures are taken, Panama will be in a good position to achieve its goals of protecting its natural resources within the Inter-Oceanic Region. *[Article copies available for a fee from The Haworth Document Delivery Service: 1-800-342-9678. E-mail address: getinfo@haworthpressinc.com <Website: http://www.haworthpressinc.com>]*

KEYWORDS. Inter-Oceanic Regional Authority (ARI), Panama, natural resource protection, land reversion

INTRODUCTION

US government land along the Panama Canal will revert to Panamanian control on December 31, 1999. In order to facilitate the land use planning and development of these lands, the Inter-Oceanic Regional Authority (ARI) created both a general and a regional plan for the area. These plans became *Law Number 21* on July 2, 1997 (ARI 1998). The general plan outlines potential land use categories, while the regional plan specifies where each category will be designated on a regional scale (ARI 1998).

This paper will use a policy science framework to examine the general and regional plans as they pertain to natural resource protection efforts in the Inter-Oceanic Region. The policy science framework is one analytical tool used to examine how decisions occur and "who gets what, how, and why" (Clark 1996). The framework "studies the process of deciding or choosing and evaluates the relevance of available knowledge for the solution of particular problems" (Lasswell 1971). The framework is broken down into social and decision-making processes. The social part of the framework examines participants based on their core beliefs, perspectives, and the methods they use for achieving results. The decision-making part of the framework is divided into six phases–estimation, initiation, selection, implementation, evaluation, and termination. At the same time, relevant past events are examined (trends), as well as why these events occurred (conditions). Based on the trends and conditions, future outcomes are predicted and recommendations are made to change the situation if the outcomes will not meet the stated goals. Using such an analysis, the paper will highlight some of the plans' potential benefits and weaknesses, and suggest alternative strategies for achieving Panama's stated goals.

BACKGROUND

In 1977, Presidents Carter and Torrijos signed the Panama Canal Treaty. This treaty was the first step to return complete control of the Panama Canal to Panama. The treaty dictates that complete reversion will occur by December 31, 1999. The economic value of the Canal and the Inter-Oceanic Region, a 10 mile wide zone surrounding the Canal, is estimated at $30 billion (Mejia, pers. comm.). The biological value is unknown, but some scientists have stated that areas in the Inter-Oceanic Region contain some of the most undisturbed forests in Central America, sheltering many endangered plants and animals (ANCON 1995, 1996a, 1996b, Strieker 1997, USAID 1997). For these reasons, the reversion represents a tremendous one-time infusion of resources into the Panamanian economy.

The Inter-Oceanic Regional Authority (ARI) was established in 1993 to "administer and promote investment" in the Inter-Oceanic Region, previously known as the Canal Zone (ARI 1998). ARI is directed by former Panamanian president Nicolas Ardito Barletta, whose appointment demonstrates the political importance of this institution. Land use maps developed by ARI designate protected wilderness areas, rural production areas, urban green areas, job creation areas, mixed use areas, residential areas, areas for uses compatible with Canal operations, and areas for other uses (S.A. Nathan Associates, Inc. 1996). The United States military will leave behind airports, schools, golf courses, ports, housing units, and forests, many of which will continue to be used by Panama (ARI 1998). The sites designated by ARI as "protected wilderness areas" are mostly forested. Some of the forests are a result of trees that regenerated at sites that were once cleared for Canal construction. These forests are now 70-90 years old. Other forest areas were never cleared, and have remained off-limits to Panamanians due to surrounding US military fences. These forests are older, although they are probably not primary forests.

SOCIAL PROCESS

One cannot analyze a problem without examining the participants involved in the process. Participants in this situation include the Inter-Oceanic Regional Authority (ARI), the Institute for the Management of Renewable Natural Resources (INRENARE), the Panama Canal Commission (PCC), the Panama Canal Authority (PCA), Panamanian citizens, the Panamanian government, and the international business community (Table 1).

The Panamanian government has at least three levels of power for its different departments. An "authority" falls in the middle, with more power

TABLE 1. A list of key participants, their values, perspectives, strategies and situation concerning the Inter-Oceanic Region, Panama.

Participant	Perspective/ Responsibilities	Situation	Base Values	Strategies	Outcomes
Inter-Oceanic Regional Authority	• Economic development is good • Development of "Canal Zone"	• Encouragement of international investment • Focus on "Canal Zone"	• Wealth • Power • Rectitude	• Economic incentives • Legislation • Marketing	• Increase wealth and power • Development of land formerly controlled by US Dept. of Defense
Institute for Renewable Natural Resources	• Protect environment • Natural resource use	• Legislative • Protection of national parks and protected areas • Community projects	• Education • Power • Respect	• Legislation • Education • Economic incentives	• Increased environmental protection in country • Increased environmental awareness
Panama Canal Commission	• Operation/Maintenance of Canal waterway	• US government organization • Responsible for Panama Canal and land immediately joining	• Power • Skill	• Fees for transit • Support from US government • Monitoring of watershed	• Operation of Panama Canal until December 31, 1999 when PCA takes over
Panama Canal Authority	• Operation/Maintenance of Canal waterway • Integration of organizations involved with Canal watershed protection	• Panama Canal and watershed area	• Power • Skill	• Don't exist yet. Probably will use incentives, legislation, education	• Increased protection of Canal watershed?
Panamanian citizens	• Interested in economic development • Multicultural • Some understanding of natural resource importance	• Varied • Rural–Indigenous, Campesinos • Urban–service, manufacturing industry	• Wealth • Well-being • Respect	• Small-scale farming • City jobs	• Clear forested land for agriculture • 50% of population lives within Inter-Oceanic Region
Panamanian government	• Want to increase economic development in country, create more stable government	• Power concentrated in a small network of wealthy elite. Often related or close social ties	• Power • Wealth • Respect	• Legislative • Diplomatic • Treaties	• Increased power and wealth among elite • Greater division between elite and poor
International business community	• Wants to increase wealth	• Being wooed by ARI to invest in development projects within Inter-Oceanic Region • Depend on Canal for transportation	• Wealth	• Economic	• Will increase their own power and wealth • Interest in maintaining Canal operation

than an "institute" but less power than a "ministry." ARI's focus is on economic development for the Inter-Oceanic Region (Mejia, pers. comm.), although it also includes natural resource protection as a guiding principle. ARI's leadership feels that the key to its success is international investment in the Inter-Oceanic Region (ARI 1998). Included in ARI's priorities are "employment generation, export promotion, development of new commercial, industrial, and tourism opportunities, development of the maritime sector, protection of the Canal Watershed, and functional and orderly urban planning" (ARI 1998).

INRENARE is the natural resource institute in Panama. It was created by constitutional law in December 1986. It's objectives are:

> To define, plan, organize, coordinate, regulate and promote the policies and activities of conservation, development, and utilization of the country's renewable natural resources, specifically the waters, soils, flora, fauna, forests, national parks, reserves, and watersheds in a form that is consistent with the national plans of development. (INRENARE 1998)

INRENARE is responsible for the protection and management of 26% of the land base, comprised of 14 national parks and 5 forest reserves, in Panama. Recent efforts to raise the status of INRENARE to an agency to give it more power and money have been successful.

The Panamanian government is another participant. It provides funding and guidance for both ARI and INRENARE. The general public of Panama is also involved in the process. Fifty percent of the country's population lives within the Inter-Oceanic Region, mostly in Panama City or Colon. While some of the general public does view environmental protection or sustainable use as worthwhile goals, others are only interested in short-term utilization (Heckadon 1993).

DECISION PROCESSES

The analytical phases of the decision process are initiation, estimation, selection, implementation, evaluation, and termination (Clark 1992). Examining what happens within each phase in Panama can help illustrate why the goals of natural resource protection in the Inter-Oceanic Region may not be achieved (Table 2).

In this case, the phases of initiation, estimation, and selection have already occurred. Initiation began after the 1977 treaty was signed and Panama started evaluating how the land reversion would affect the country. Estimation began when ARI was created and started investigating possibilities for land use within the Inter-Oceanic Region. The general and regional land-use

TABLE 2. Decision process phases and potential pitfalls, including assessment and specific examples from Panama.

Phases	Potential Pitfalls	General Assessment	Example
Initiation– *problem identification* • What can Panama do to make the most of the reversion of the US military lands in the former Canal Zone?	• Letting a single agency dominate • Defining the problem as simply as possible with a single objective • Looking only to the short-term	• ARI is in control • Objectives were both economic development and natural resource protection	• Goal statements by ARI
Estimation– *expert analysis* • What information is needed?	• Not asking the hard questions • Exaggerating the expected benefits or discounting the possible costs	• ARI looked outside itself for technical advice on planning	• ARI brought in consulting groups from US to help it develop a land-use plan • TNC conducted rapid ecological assessment on US military lands. Was this information used?
Selection– *creating a plan* • What should Panama choose as a land-use plan for the Inter-Oceanic Region?	• Not coordinating governmental decision making, particularly when there are stronger and weaker agencies	• This may have been a problem between ARI and INRENARE. Both are very focused on their power (or lack their of) within government	• General and Regional plans were made into law in 1997
Implementation– *applying a specific* *program* • How will ARI's general and regional plans be put into place on the ground?	• Not having adequate coordination or organization between participants	• Problems are developing. This is the stage at which Panama is now • Lack of attention to details about how to implement plans–not clear who will manage protected areas or what the goals of those areas will be	• Confusion over who will be responsible for management of protected areas. ARI representative said they would be responsible until 2005, at which point INRENARE would take over. Others said INRENARE would be responsible in conjunction with PCA
Evaluation– *appraisal of program* • Are the goals of economic development and natural resource protection being met?	• Not being sensitive to criticism • Ignoring past experiences, repeating same mistakes	• There is no publicized established evaluation process	• At least two areas should be considered for evaluation–biophysical (what is happening within the watershed) and organizational (are participants helping or hindering the achievement of the stated goals)

Phases	Potential Pitfalls	General Assessment	Example
Termination–completing/ending a program • When are ARI's regional and general plans finished?	• Assuming that program termination mean failure	• There is no publicized established termination process	• When ARI was created they were also designed to disassemble in 2005. How will this happen?

plans were developed, and ARI has begun implementing them. Problems are arising because there is a lack of attention to the details surrounding the implementation of these plans for protected areas within the Inter-Oceanic Region. This has enormous potential to prevent Panama from achieving its goal of protecting the natural resources in the area.

Specifically, there needs to be clarification about which organization will be managing the newly-designated protected areas. In talking to ARI officials, they said ARI will be managing this region until 2005, at which point INRENARE will take over (Mejia, pers. comm.). Others said that it will be INRENARE in conjunction with the PCA (Lesnick, pers. comm.). Among the participants we interviewed during our limited time in Panama, there was not a consensus as to exactly how the management will be implemented. ARI was established as a development and planning organization, and its expertise is in attracting foreign investment to Panama. It is unclear whether its employees have experience with natural resource management. INRENARE is already responsible for management of 26% of the Panamanian land base and its funds are limited (Endara, pers. comm.). If INRENARE is the responsible agency, will its budget increase so that the organization can allot the time and effort to properly manage these areas? Does INRENARE have the staff capability necessary to meet the management goals for the protected areas?

ARI's plans also need to clarify management goals for protected areas. In the general plan, ARI lists the following as purposes for "protected wilderness areas":

- Meet the need to regulate the hydrological cycle, thus guaranteeing the operation of the Canal and preserving the regional biodiversity.
- Provide green areas near the population centers.
- Protection of the ecosystems (ARI 1998).

ARI specifies that the designation "protected wilderness areas" includes protective forests, protected landscape areas, hydrologic protection zones, and recreation areas. However, details are lacking as to what exactly may be done on these lands. For example, people live in many of Panama's national parks–whether people will be allowed to move into some of these new protected areas is unclear. Development of tourism on and near these protected areas is being encouraged, but again it is not clear how much this development will be allowed to impact the protected areas. If these questions are not answered before the land reverts, Panama will encounter serious difficulties with the implementation of its plans and will not be able to meet its goal for natural resource protection in the Inter-Oceanic Region.

There is also little information regarding evaluation of the plans. If no formal evaluation process has been established, there will be no way to know if the plans' goals have been met. There are at least two aspects that should be evaluated–biophysical changes and organizational operations. Biophysical changes could be monitored in a number of ways, including amount of forest cover, number of people in the area, or frequency of fire. These evaluations would help determine if the protected areas are truly being protected. It would also be helpful to evaluate participating organizations, to see if they are helping or hindering the fulfillment of the plans' stated goals. It is only when evaluations occur that it is possible to learn from previous mistakes. The lack of attention to details within the implementation and evaluation processes illustrates areas where ARI's plans can be improved in order to meet the organization's goals for natural resource protection in the Inter-Oceanic Region.

ANALYSIS

Trends and conditions are the "what and why" of the decision process. Trends describe events in the past which have led to the current situation, while conditions are the factors that influenced the direction and intensity of these trends (Clark 1992). The three trends and conditions that have played particular importance in the development of plans for the Inter-Oceanic Region in Panama are described below (Table 3).

While Panama is doing quite well compared to some of its neighbors, there is still a desire for greater economic growth and stability. Monetary resources are limited. The government's goals include strengthening the country's economy, creating a politically stable government, and raising the living standard (Anonymous, 1998). Using funds to accomplish these goals constrains its funding for environmental issues and enforcement of existing laws and regulations that would protect the environment (Anonymous 1998). INRENARE has an annual budget of about $17 million dollars (Endara, pers. comm.). With this money, they must manage and protect 26% of the country's land base. In Soberania National Park, which lies within the Panama Canal Watershed Areas, funding constraints dictate that there are only ten park guards responsible for patrolling, maintaining, and educating throughout the entire 22,104 ha of the park (Ramos, pers. comm.). If natural resources continue to be viewed as unlimited and more emphasis is not placed

TABLE 3. Trends that affect ARI's plans in the Inter-Oceanic Region, conditions affecting trends, projections if those trends remain, and recommendations to modify future projections.

Trend	Condition	Outcome/Projections	Recommendations
• Government wants to develop economy, create more stable government, raise living standard	• Expectations of participants that government will help them increase their wealth	• Increase participants in social process • Decrease resources (because of allocation of funds to these priorities, less money to fund environmental protection and enforce environmental regulation)	• Shift perspectives to sustainable development • Manage for a longer time period • Increase funds for natural resource protection (i.e., by raising toll on Canal)
• Belief that Canal operation is key to economic development of region	• History has shown this to be true. Previous protection of Canal watershed has been able to maintain waterway	• Continued concern for and protection of natural resource within the watershed • Concern may increase if 3rd lock is opened	• Find ways to accelerate trends and conditions
• Large, broad based environmental goals (Constitution, ARI's goals)	• People understand value of natural resources within country • Understand connection between maintaining Canal operations and protection of watershed	• Greater/better protection for natural resources of country	• Find ways to accelerate trends and conditions

on their protection, then the outcome will be a decrease in the natural resource base of Panama (Heckadon 1993).

Top officials in the Panamanian government and ARI understand that maintaining boat traffic through the Canal is the key to economic development in the region. This understanding is demonstrated in one of ARI's guiding principles for its development plans:

> To provide for the long term preservation of the natural resources necessary for the operation of the Panama Canal, the economic hub of the Inter-Oceanic Region, emphasizing the protection of hydrologic resources and the prevention of environmental degradation, such as erosion, which may affect the efficient operation of the Inter-Oceanic waterway and its potential expansion in the future. (ARI 1998)

Any development that occurs within the Inter-Oceanic Region will not survive if the Canal cannot operate. Commercial development depends on boats being able to pass through the Canal and the capability to load or unload goods at these sites. Tourism development relies on interest about the Canal. If the Canal is not operational, visitors will not be as likely to come. Officials realize that in order to maintain the waterway, they must prevent sediment from entering the Canal and they must maintain a sufficient supply of water to the Canal.

To ensure that the Canal remains operational, Soberania and Chagres National Parks were created to protect critical areas in the Canal Watershed. Areas within the Inter-Oceanic Region have also been designated as hydrologic protection zones in order to prevent erosion into the Canal (ARI 1998). The government's recognition of this has led to continued concern for and protection of the natural resources within the Watershed.

Panama has a history of large, broad-based environmental goals and laws. This dates back to the country's Constitution, part of which states:

> It is a fundamental duty of the State to guarantee that the living population have a healthy environment free of pollution, where the air, water, and food satisfy the requirement of adequate development for human life (Article 114). The State and its national territorial inhabitants are obligated to sponsor social and economic development that prevents the polluting of the environment, maintains ecological balance and prevents the destruction of ecosystems (Article 115). (Valiente 1998)

Overarching environmental goals are also included in ARI's regional plan. The following is listed as one of the guiding principles:

> To provide an opportunity for demographic, economic, and urban growth to take place in the Inter-Oceanic Region during the next 25 years based on the on-going dynamics of such growth, while preserving the region's natural resources wealth and its potential. (ARI 1998)

The government understands the value of the country's natural resources. Since this realization exists, there is likely to be greater protection for the natural resources of the country than would otherwise occur. However, this understanding conflicts with the first trend of wanting to develop the economy and raise the standard of living, and so there will be a continued struggle to balance all of these goals.

RECOMMENDATIONS

Alternatives to ARI's plans exist for natural resource protection within the Inter-Oceanic Region. One alternative is to do absolutely nothing. Panama has not chosen this option, and is taking a role in protecting and managing the natural wealth it possesses. However, these efforts are not enough. A second alternative is to maintain the status quo, to keep doing everything that Panama is currently doing. A third option is to expand current efforts so that the plans' weaknesses are eliminated. It is only with an expansion of the status quo that Panama will be able to achieve its goals for natural resource protection within the Inter-Oceanic Region. The following are just two areas where efforts spent will dramatically increase the likelihood of the attainment of these goals.

Increase the Specificity of the Regional Plan

ARI's regional and general plans contain unclear concepts about how natural resource protection will happen within the Inter-Oceanic Region. If the details surrounding implementation of the plans remain vague, the country will not meet its objectives for maintaining its natural resources. In particular, ARI needs to designate a specific organization to manage the new protected areas. Clear goals for management should be defined. The purpose of protected wilderness areas listed in the regional plan is to "meet the need to regulate the hydrological cycle, thus guaranteeing the operation of the Canal and preserving the regional biodiversity" (ARI 1998). Protected areas are designed to "preserve the existing forests and other natural areas as a green area that could lead to a transition towards protective forest designation" (ARI 1998). These goals still leave many unanswered questions, which makes management of these areas extremely difficult. For example, do these goals allow for people to move into a protected area? What if these people were to practice best management techniques that would reduce soil losses? Questions like these need to be answered now or there will be confusion and possibly failure in the future.

Establish Evaluation Procedures for Biophysical Changes and Organizational Behavior

If there is no established evaluation process, it will be difficult to determine if the protected areas are actually maintaining and protecting the natural resources within them. There are a variety of research organizations within the area (e.g., Smithsonian Tropical Research Institute and Panamanian universities) that could help both with the design and execution of biophysical monitoring efforts. Monitoring alone is not enough—the results from monitoring need to be compared with the stated goals of the plans. Monitoring tools could include the use of ground-truthing by aerial photos and park guards. Aerial photos taken over time would show any changes in forest cover. Because they work in the protected areas, park guards are a resource that could be used to help confirm changes in human occupancy of the area. They could also help examine the incidence of fire in these regions, as well as ascertain whether the fires were in support of agriculture or were set maliciously. Without monitoring efforts, it will be difficult to assess the status of protected areas and whether the stated goals for them are being met.

Evaluations of organizational function are needed to determine if an involved agency helps or hinders the attainment of the goals. If the organization is promoting the goals of the regional plan, then it should be commended and its management strategies used as a model. If the organization is not helping, then problems should be identified and addressed. This evaluation could also examine whether particular participating agencies even have the capability to manage these lands. If they do not, training programs need to be created to increase institutional capability. These programs could focus on leadership training, problem-solving exercises, and staff development. At the same time, the organizations should provide incentives and support for employees who want to expand their own education relating to natural resource management.

CONCLUSION

Within the next year and a half, Panama will undergo tremendous changes. The country has begun preparations for these changes by establishing ARI and developing plans for specific land uses within the Inter-Oceanic Region. ARI's regional plan is a huge step in the right direction necessary to develop the area and protect the country's natural environment within the Inter-Oceanic Region. While Panama has created a model that tries to balance economic development and natural resource protection, the plans lack clarity and specificity about how the natural resource protection efforts will be implemented and evaluated. In particular, the plans should be specific about who will manage the new protected areas and what the management goals for these areas will be. An evaluation mechanism should be established to determine if the plans are achieving these goals. There is still time to clarify and enhance current efforts. If the natural resource aspects of the plans are made more specific, Panama is in a good position to accomplish its goals for natural resource protection within the Inter-Oceanic Region.

REFERENCES

ANCON. 1995. *Evaluacion ecologica de la cuenca hydrografica del canal de Panama.* National Association for the Conservation of Nature, Panama City, Panama.

ANCON. 1996a. *Ecological survey of US Dept. of Defense lands in Panama. Phase III: HOROKO, Empire Range, and Balboa West Range.* The Nature Conservancy, Arlington, VA.

ANCON. 1996b. *Ecological survey of US Dept. of Defense lands in Panama. Phase IV: Fort Sherman, Pina Range, and Naval Security Group Activity, Galeta Island.* The Nature Conservancy, Arlington, VA.

Anonymous. "Environmental Issues in US Military Host Nations." Online. Internet. 31 March 1998. Available http://aepi.gatech.edu/etrens94/etr_15.html. Arauz, Alejandra. 1997. "ANCON: The fight to save the Panama Canal watershed." *The Isthmian.* Online. Internet. 30 March 1998. Available http://www.isthmian.com/nov97/ecology/ancon.html.

ARI. *Memoria Annual 1996-1997.* Online. Internet. 31 March 1998. Available http://www.ari-panama.com/.

ARI. Annex I & II. Online. Internet. 31 March 1998. Available http://www.ari-panama.com/.

Clark, T.W. 1992. Practicing Natural Resource Management with a Policy Orientation. *Environmental Management.* 16(4): 423-433.

Clark, T.W. 1996. Policy processes and conservation biology. In S. Bessinger, R. Lacy, A. Lugo, and T. Clark (eds.). *Conservation Biology: From theory to practice.* Oxford University Press, New York.

Endara, M. Director General of INRENARE. Discussion of natural resource management in Panama. Personal communication. March, 1998. Gamboa, Panama.

Heckadon Moreno, S. 1993. Impact of development on the Panama Canal environment. *Journal of Interamerican Studies and World Affairs.* 35(3): 129-149.

INRENARE. Online. Internet. 30 March 1998. Available http://www2.usma.ac.pa/~eco1/inrenare.htm

Lasswell, H.D. 1971. *A Pre-View of the Policy Sciences.* American Elsevier Publishing Company, Inc. New York.

Lesnick, B. Consultant, Executive Service Corps. Discussion of natural resource management in Inter-Oceanic Region, Panama. Personal communication. March 1998. Stamford, CT.

Mejia, Alfredo. Representative, Inter-Oceanic Regional Authority. Discussion of ARI's role in Inter-Oceanic Region, Panama. Personal communication. March, 1998. Gamboa, Panama.

Ramos, T. Guard, Soberania National Park, Panama. Discussion of guard's duties in park. Personal communication. March, 1998. Panama City, Panama.

S.A. Nathan Associates, Inc. 1996. "Mapa °2, Plan General de Uso del Suelo del Area del Canal Region InterOceanica." Online. Internet. 29 March 1998. Available http://www.ari-panama.com/.

Strieker, G. 1997. "Forests along Panama Canal face uncertain future." Cable News Network. Online. Internet. 2 April 1998. Available http://forests.lic.wisc.edu/gopher/centamer/panforest.txt.

Valiente, Arecio. "Do indigenous people have the right to decide about their own natural resources?" Online. Internet. 30 March 1998. Available http://ecocouncil.ac.cr/indig/conventi/panam-eng.htm.

USAID Congressional Presentation FY 1997. "Panama." Online. Internet. 30 March 1998. Available http://www.info.usaid.gov/pubs/cp97/countries/pa.htm.

Balancing Conservation and Economics: The Development of an Ecotourism Plan for Panama

Katherine Lieberknecht
Jennifer Papazian
Andrea McQuay

SUMMARY. Ecotourism has been used successfully in many countries to promote economic well-being, conserve natural resources, and promote community development. A thorough examination of the factors that make a successful and unsuccessful ecotourism program was conducted in this paper. This analysis was performed in order to formulate recommendations for the development of an ecotourism program in the Panama Canal Watershed that will meet these objectives. A careful look at the conservation of natural areas, community development, and the

Katherine Lieberknecht, Jennifer Papazian, and Andrea McQuay obtained Master of Environmental Studies degree from the Yale School of Forestry and Environmental Studies, New Haven, CT 06511.

Katherine Lieberknecht is now in landuse planning with the Finger Lakes Land Trust, Ithaca, NY.

Jennifer Papazian is Environmental Consultant in Southboro, MA.

Andrea McQuay is Specialist in environmental education for the Tennessee Chapter of the Nature Conservancy, Nashville, TN.

The authors would like to thank the valued contributions and editing skills of Mark S. Ashton, Luisa del Carmen Camara, Christopher Elwell, Bruce Hammond, Robert Hauff, and Jennifer O'Hara. The authors would also like to thank everyone who spoke with them, took them on tours, and hosted them while they were in Panama.

[Haworth co-indexing entry note]: "Balancing Conservation and Economics: The Development of an Ecotourism Plan for Panama." Lieberknecht, Katherine, Jennifer Papazian, and Andrea McQuay. Co-published simultaneously in *Journal of Sustainable Forestry* (Food Products Press, an imprint of The Haworth Press, Inc.) Vol. 8, No. 3/4, 1999, pp. 107-126; and: *Protecting Watershed Areas: Case of the Panama Canal* (ed: Mark S. Ashton, Jennifer L. O'Hara, and Robert D. Hauff) Food Products Press, an imprint of The Haworth Press, Inc., 1999, pp. 107-126. Single or multiple copies of this article are available for a fee from The Haworth Document Delivery Service [1-800-342-9678, 9:00 a.m. - 5:00 p.m. (EST). E-mail address: getinfo@haworthpressinc.com].

economics of ecotourism fleshed out the good and the bad of existing programs. Case studies from other developing countries were used in order to set up criteria that should be used in Panama's blossoming ecotourism industry. Panama has a vast array of natural and cultural resources that can benefit from a carefully planned ecotourism program. When planning this initiative it is very important to take the politics of the country and communities into consideration, make sure the project is scaled to the carrying capacity of the natural resources and community involved, use existing infrastructure, train a skilled local work force, and maintain monitoring programs that ensure that an unsuccessful program be either improved or terminated. If these steps are taken, an ecotourism program in Panama could be very successful. *[Article copies available for a fee from The Haworth Document Delivery Service: 1-800-342-9678. E-mail address: getinfo@haworthpressinc.com <Website: http://www.haworthpressinc.com>]*

KEYWORDS. Conservation, community development, ecotourism, national parks, Panama, protected areas

INTRODUCTION

The purpose of this paper is to explore the potential for ecotourism in the Panama Canal Watershed. First, we examine Panama's unique natural, cultural, and economic characteristics that make it an ideal location for an expanded tourism industry. The term ecotourism is defined, and the three goals of successful ecotourism are explored: achievement of conservation objectives, enhancement of community development, and generation of a national economic benefit. Lastly, we present recommendations for the establishment of a beneficial ecotourism program in Panama.

The Panama Canal Watershed possesses a rich natural and cultural heritage. Protected areas, including four national parks and the famous Barro Colorado National Monument, encompass almost 370,000 acres of the Watershed. The diverse landscapes of these tropical forests are home to thousands of plant and animal species, including 230 mammal species, 169 amphibian species, and 929 bird species (Endara 1997). Bird watching and hiking are just a few of the diverse activities to enjoy within the park system. The Watershed has coastline on both the Atlantic and Pacific coasts, where record-setting fishing exists, as well as coral reef diving. Lake Gatun, the world's largest human-created lake, provides an area for water recreation, such as boating and fishing (Tribaldos 1997).

Amongst all the natural beauty of Panama lies a fascinating cultural history. Panama's central position as a transit zone has given the country a considerable amount of ethnic diversity (Haxton 1995). City streets pulsate with Panamanians of Spanish, Greek, Caribbean, and Japanese descent, among others. The Watershed contains enclaves of indigenous tribes such as the Embera, and the Kuna tribal members sell their world-renown *molas* in open air markets. Colonial ruins of Old Panama and the submerged villages in Lake Gatun serve as reminders of the past. Colon and Panama City not only offer a multi-cultural atmosphere, but also shopping in the tax-free zone and sight-seeing of the canal. These features offer a tremendous potential for the development of a tourism program.

Panama's economy is making significant strides towards economic growth by gradually lifting its protectionist barriers and promoting economic diversification. Despite these efforts, unemployment remains high and is projected to continue on this trend (Ministry of Labor and Social Welfare 1996). Thus the challenge is to create new jobs and to develop the "human capital" to support the newly expanded sectors. Policies to promote investments, lower production costs, and improve the efficiency of services and infrastructure are underway. Many believe that the opportunities accompanying the administration and management of the canal and surrounding land entitlements will stimulate economic growth especially in terms of infrastructure (Autoridad de la Region Interoceanica 1997a).

By 2000 the Republic of Panama will receive US military bases comprising 28,000 ha, and the proper Canal operation area, another 27,000 ha essential for canal functioning. Additionally, there are 1,600 ha dedicated to tropical research. Currently development plans initiated by the government, foreign investors and private multi-national corporations have proposed everything from tourism development, private clubs, golf courses, to liquid gas storage units and container storage and repair centers (Autoridad de la Region Interoceanica 1997b). The canal area has great potential for infrastructure development both on the Atlantic (13,800 ha) and Pacific (16,000 ha). There is debate, however, over the management of natural resources that exist or had once existed throughout the canal area. Several environmental groups have put forth proposals to reforest and conserve large tracts of forest within the canal. Their struggle against a growing and expanding nation and enthusiastic foreign investors will be difficult.

The reversion of the Panama Canal and its surrounding lands presents Panama with an extraordinary resource, both from an ecological and economic standpoint. The Canal watershed has a remarkable amount of forested land that supports incredible biodiversity and provides the Canal with water. It also represents a huge economic opportunity for Panamanians. Economic activity that is compatible with conservation of these forest resources is needed. One option is ecotourism. In order to meet the specific objectives of an ecotourism venture, the program must be designed specifically for the target area, and should be scaled on a level that is appropriate to each individual project.

ECOTOURISM AND THE CONSERVATION OF NATURAL AREAS

The goal of ecotourism is to capture a portion of the enormous global tourism market by attracting visitors to natural areas and using the revenues to fund local conservation and fuel economic development. (Ziffer 1995)

Much debate has been given to the definition of "ecotourism" and to whether or not it should apply to general nature tourism or to a more specific type of tourism (Ceballos-Lascurain 1993, Fennel and Eagles 1990). The word ecotourism was first used in 1991 by Hector Ceballos-Lascurain, who defined it as "travel in undisturbed, natural areas with the objective of admiring, studying, and enjoying the scenery and its wild animals and plants and culture." This definition has evolved to encompass both environmental and cultural goals. A current definition of ecotourism is "purposeful travel to natural areas to understand the cultural and natural history of the environment; taking care not to alter the integrity of the ecosystem; producing economic opportunities that make the conservation of natural resources beneficial to local people" (Ecotourism Society 1993, 1998). "Nature-based tourism" is generally seen as tourism focused on nature, but on a scale large enough that ecosystem integrity is likely altered (Brandon 1996).

As indicated in one of the many definitions that can be found to describe ecotourism, conservation is and should be one of the three major goals of any ecotourism program. Because of this, the concept of ecotourism has been embraced by many environmentalists, conservationists and wildlife managers looking for a quick solution to the continued degradation of the world's natural resources. The idea has also become very popular with developers, governments, and individuals of many cultures, due to the economic and social incentives that accompany the industry (Norris 1992). For this very reason the use of ecotourism needs to be managed carefully in order to maintain the original purposes of this new type of tourism.

A rigorous definition of ecotourism, as defined by the Ecotourism Society, emphasizes the responsibility to conserve natural environments and the needs and well-being of the local people of an area through responsible travel (Ecotourism Society 1993, 1998). Ideally this allows developing countries the opportunity to capitalize on their natural attractions without incurring the adverse effects now commonly associated with conventional tourism. Unfortunately, poor management often allows too much encroachment on natural areas, furthering degradation, while at the same time increasing funding to maintain the area (Wheat 1994). These two objectives do not need to be mutually exclusive, but in many cases they appear to be. Therefore, the question needs to be asked whether ecotourism can be successfully implemented as a benefit to the world's resources and indigenous cultures, rather than as just another tool that destroys our natural resources for profit.

There are cases for both sides (see Table 1). A successful ecotourism story can be found in the Virunga Mountains of Rwanda. Farmers clearing land higher and higher up the slopes of the mountains, deforested large amounts of mountain gorilla habitat decreasing the population size, while at the same time destroying the very watershed that they relied on for agriculture (Arnold 1995). The implementation of the Parc National des Volcans in 1979 was the start of a conservation goal that would serve to both increase foreign capital through tourism and provide an economic reason for Rwanda to protect the mountain gorilla (Foster 1992). This initiative was successful on all fronts. Some of the reasons for this include: local participation, keeping a large percentage of income at the local level, local people were included in Gorilla conservation and were able to observe the benefits of ecotourism first hand, and having a focus species that tourists were almost guaranteed to be able to see (Brandon 1996). Up until the start of the Civil War in Rwanda, tourism was Rwanda's major industry and the gorilla's had become a source of local pride. Most importantly, gorilla populations had been steadily recovering under this program, along with the continued protection of the watershed (Reed 1995).

Not only is this project an excellent example of an ecotourism success story, it is also an excellent example of how, without political stability, even the most successful project is doomed to failure. With the start of the Rwandan Civil War, all tourism to Rwanda was halted. In addition, poaching of mountain gorilla's, and gorilla's caught in the crossfire of civil war, has again ravaged the population, and negated ecotourism's earlier success (Reed 1995). Thus, in the designing of an ecotourism program, it is very important to take the country's political climate and stability into account, before setting up a program that may not be politically feasible.

Another successful ecotourism program can be found at the Monteverde Cloud Forest of Costa Rica. Although a very popular attraction that could easily be overcrowded and degraded by too many tourists, this reserve has maintained a strict conservation objective. Monteverde limits the amount of visitors on the trails at all times and maintains the wild experience for all visitors and wildlife residents (Norris 1994).

Unfortunately, there are some programs that have not been as successful in regards to conservation, as the Mountain Gorilla Project and Monteverde. One such example is Manuel Antonio National Park in Costa Rica. It is the most popular, most visited national park in Costa Rica (McNeil 1996). Be-

TABLE 1. Useful ecotourism studies.[1]

Citation	Location of Project	Ecotourism Activity
Ayala, H. 1997. Resort Ecotourism: a catalyst for national and regional partnerships. *The Cornell Hotel and Restaurant Administration Quarterly* 38(4): 34-45.	Not focused on a specific country or region.	Guidelines for how to establish an economically successful and environmentally sustainable ecotourism resort venture.
Brandon, K. 1993. Basic steps in encouraging local participation in nature tourism projects in ecotourism: a guide for planners and managers. The Ecotourism Society, Alexandria, Virginia.	Not focused on a specific country or region.	Informational guide tailored to expanding local participation in ecotourism project implementation.
Brandon, K. 1996 Ecotourism conservation: A review of key issues. The World Bank. Paper No. 33.	Not focused on a specific country or region.	Excellent overview of the history and future of ecotourism, as well as its positive and negative aspects.
Church, P. and K. Brandon. 1995. Strategic approaches to stemming the loss of biological diversity. United States Agency for International Development, Center for Development Information and Evaluation, Washington, DC.	Not focused on a specific country or region.	Ecotourism is offered as one component of a plan for reducing world-wide biological diversity loss.
Ecotourism Society. 1998. Ecotourism Society's web page: www.ecotourism.org.	Various countries and regions.	Excellent overview of ecotourism guidelines, research, current projects, and references.
Foster, J.W. 1992. Economic considerations in mountain gorilla conservation. In: pp. 87-91. Proceedings Joint Meeting AAZV/AAWV.	Rwanda.	Overview of the economic basis of mountain gorilla conservation in Rwanda.
Lindberg, K. 1991. Policies for maximizing nature tourism's ecological and economic benefits. The Ecotourism Society, Washington, DC.	Not focused on a specific country or region.	Guidelines for balancing and maximizing the ecological and economic benefits of ecotourism projects.

Citation	Location of Project	Ecotourism Activity
Meadows, D. 1995 Beyond Shamans, Toucans, and Tourist: Local Participation in Ecotourism in Ecuador and Costa Rica. Proceedings of the 1995 meeting of the Latin American Studies Association.	Ecuador and Costa Rica.	Specific case studies of local participation in ecotourism projects in two Latin American countries.
Miller, J. and E. Malek-Zadeh, editors. 1996. The Ecotourism Equation: Measuring the Impacts. Yale School of Forestry and Environmental Studies Bulletin 99.	Articles include many site locations, with a strong focus on Sub-Saharan Africa and Latin America.	Proceedings on an ecotourism conference. Articles encompass many ecotourism topics and case studies.
Norris, R. 1992. It's green. It's trendy. Can ecotourism save natural areas? *National Parks* 66(1-2): 30-34.	Several case studies in Latin America.	Short critique of how ecotourism has lived up to conservation expectations.
Norris, R. 1994. Ecotourism in the national parks of Latin America. *National Parks* 68:33-37.	Latin America.	Overview of ecotourism successes and failures within National Parks in Latin America.
Reed, S. 1995. Survivors in the mist. *People* March 6:40-45.	Rwanda.	News magazine article on the Rwandan gorilla ecotourism project.
Taylor, M.R. 1996. Go, but go softly. *Wildlife Conservation* March-April: 12-19.	Not focused on a specific country or region.	Overview of limitations of ecotourism.
Wells, M. and K. Brandon. 1992. People and Parks: Linking Protected Area Management with Local Communities. World Bank.	Most case studies are focused on Latin America.	Publication is not solely focused on ecotourism, but presents some ecotourism case studies and guidelines, especially about local participation.
World Tourism and Travel Organization (WTTC). 1995. Travel and tourism's economic perspective. A special report from WTTC.	Not focused on a specific country or region.	Excellent economic overview of ecotourism.

[1] This list is not an exhaustive list of all ecotourism studies, but rather a summary of studies the authors found particularly useful.

cause of this, however, visitors find crowded trails, degraded jungle, and white-faced monkeys that have been pushed out of the jungle and out onto the beach in order to beg tourists for food. This is certainly not the conservation success story that most environmentalists are looking for.

There are many other examples of overcrowding and degradation that are similar to Manuel Antonio. These examples are in no means meant to condemn Costa Rica's excellent national park system or their innovative and successful approach to tourism. They are only meant to point out a few of the pitfalls that can be encountered with ecotourism programs, and be used to in order to make suggestions that will hopefully help fledgling programs avoid these issues.

In theory, ecotourism is an excellent solution to increased tourists and increased degradation to natural areas. However, ecotourism inevitably has the same expansion potential as traditional tourism, which will require more space for tourists and increased clearing of land. It cannot automatically be assumed that the "ecotourist" is environmentally sensitive. Careful attention needs to paid the carrying capacity of the different areas, so that this capacity is not exceeded, thereby destroying the sustainability of the program. Conservation and community development should not be allowed to be usurped by unbridled economic gain.

Panama has a great opportunity to preserve some of its beautiful natural resources in the Canal Watershed. One way of approaching this is with the implementation of an ecotourism program. However, this program should be planned carefully, taking other programs into consideration, establishing an ecotourism plan as a guide that can be administered and monitored by a group such as Panama's National Institute of Renewable Natural Resources (INRENARE), and by making sure that all plans are adaptive and can be modified as necessary. In addition, a carefully planned program that accounts for not only conservation, but includes a community development component, and attention to economic gains, could potentially benefit all participants in this process.

COMMUNITY-BASED ECOTOURISM

Ecotourism is most simply defined as sustainable tourism. According to Sproule (1996), to be considered sustainable, projects must fulfill three broad criteria: (1) financial support for protection and management of natural resources; (2) economic benefits for residents living near natural areas; and (3) encouragement of conservation among these residents, in part through economic benefits. With its vibrant cultural history and copious biological diversity, Panama has the potential to become the premier ecotourist destination. However, to remain sustainable during and after the inevitable growth spurt the country will undoubtedly experience after the rights to the Canal are conferred to it, will require an extensive amount of planning and innovation. One approach to building an ecotourism industry that intends to provide the benefits to the most people while helping to conserve the natural environment is community-based ecotourism. Community-based ecotourism is committed to an integrated conservation and development project (ICDP) method. ICDP's "aim to achieve protected area conservation by promoting socio-economic development and providing local people with alternative income sources which do not threaten to deplete the flora and fauna of the protected area (Brandon 1996)." Such projects can include biosphere reserves, multiple-use planning areas, buffer-zones, regional conservation areas and sustainable eco-tourism (Brandon 1996).

Ecotourism ventures considered to be community-based are those that are managed and/or owned by the local community. The sustainability of com-munity-based ecotourism enterprises is further determined by the willingness of the community to care for its natural resources in order to generate income and subsequently dedicating this income toward bettering the lives of its members. This duality of outcome suggests that associated with any commu-nity-based entrepreneurial endeavor there are specific gradations of partici-pants and beneficiaries (Wells and Brandon 1992). To maximize the amount of those participating and benefiting from ecotourism at the local level, successful community-based initiatives combine elements of conservation, busi-ness enterprise and community development. Within and among these strate-gies there are often myriad conflicts that arise and ultimately doom the project to failure. Thus it is essential that particularly salient aspects of the community are considered before the ecotourism project is initiated. These aspects include but are not limited to: social structure of the community (identity), relationship with the land (property rights), projected level of participation, and applicability of forming substantive partnerships with exte-rior entities (Brandon and Margoluis 1996; Sproule 1996).

The most successful community-based ecotourism ventures have devel-oped in communities with a cohesive social structure (Brandon 1996). Com-munities that are secure with their identities and share common visions and projections for the future are perhaps best equipped to deal with the imminent changes that will occur when the community is suddenly opened up to outsid-ers. Cohesive communities can exercise greater control over the development of tourism and spread of impacts. Additionally, benefits can be shared more equitably across the community when the social structure is well defined and understood.

The community's relationship with the land seriously impacts the success of ecotourism. A certain degree of autonomy over the land supplemented with a sense of stewardship and long-term conservation is key to making any enterprise that depends on a resource work well (Meadows 1995). Ownership tends to imply a greater sense of responsibility toward sustainably managing land. Thus the coupling of land ownership with ownership of (or active partnership in) the tourism investment can help to bolster the chance of success (Wells and Brandon 1992).

There are several levels of community participation that should be considered when instituting an ecotourism plan. Not every community should be encouraged to own and manage its venture. In some instances it is advisable to pursue alternative arrangements such as partnerships and external management coordinators. Communities that lack strong local institutions or do not have a cohesive social structure are perhaps better suited for ecotourism ventures that partner them with other organizations such as non-governmental organizations (ANCON), universities (University of Panama), research organizations (STRI) or government agencies (INRENARE). These institutions can assist the community in developing strategies to deal with the types of issues that arise out of commercial ventures and may also have capital to invest in the project. It is important to monitor the outcomes of such partnerships to ensure that the interests of the community are not subjugated to the outside partner. Empowerment of the local community, both economically and environmentally, is one of the greatest benefits of community-based ecotourism projects, and it should not be sacrificed to the development objectives of external interests. The Kuna Indians provide an excellent example of a mutually beneficial partnership forged between indigenous groups and the private sector (Brandon 1996).

The primary goal of community-based ecotourism in Panama should be to encourage the conservation of natural resources by integrating conservation and development. Fulfillment of this goal depends on how the benefits of the venture are perceived by the community. The short-term costs of foregoing the resource must be met, at least to some degree, by benefits accrued from the project. These benefits must in-turn be clearly seen as derived from and dependent upon the resource base and thus its sustainable management. Benefits must be shared equitably throughout the community and preferentially equated to level of participation. And, the local institution should in the process be strengthened so to increase their capacity for collective resource management (Sproule 1996).

There are several alternatives for building successful community-based ecotourism ventures in Panama. First, it would be useful to create prototypes in communities that are identified as having the essential characteristics described above. Constructing these demonstration projects and then inviting communities with similar potential to visit and participate for a day would help provide communities with the ideas and useful information and provide a positive role model. Second, workshops and continuing education classes would be useful in supplementing and refreshing the information gained through visiting the demonstration sites. These programs could help in creating a network of shared experiences between participants and those wishing to start-up new projects and those trying to reinvigorate existing ones. Lastly, continued monitoring of these community-based projects is useful in evaluating the success in meeting stated goals and objectives.

ECOTOURISM ON A NATIONAL SCALE

Panama's Current Tourism Plans

Panama's government highlighted tourism as a national priority for economic development in 1992 (Government of Panama 1992). This mirrors a shared desire of all Central American nations to make tourism the number one source of income for the region (WTO 1996). As a first step, Panama and the Organization of the American States (OAS) forged a technical-cooperation agreement, which laid the groundwork for Panama's master plan of tourism (*Plan Maestro de Desarrollo Turistico de Panama*) (Panamanian Institute of Tourism (IPT) and OAS 1993). This plan is one of the first national tourism strategies centered on natural and cultural based ecotourism (Ayala 1997a). The basic goal of Panama's master plan is to establish a set number of accommodations and services for each of Panama's attractions. Panama hopes to achieve this by promoting the country's cultural and natural attributes, mainly by attracting foreign investment, which is hoped will stimulate the national economy (Ayala 1997b).

Panama is encouraging tourism in two ways: by providing economic incentives and by promoting Panamanian tourism through marketing. In 1994, the Panamanian government created Law 8 to promote tourism development (Government of Panama 1996). Law 8 provides tax incentives and benefits for the tourism projects in Panama. These include: tax exemptions for up to 20 years on materials, boats, vehicles, and equipment; 20 years of exemption of real estate tax; tax exoneration over the investor's capital; 20 years exoneration from income tax on interest charged by creditors in hotel investments; and real estate depreciation at 10% per year.

Combined with these regulations, Panama is hoping that promotion will create a competitive climate for tourism. Panama's main vehicle of promotion is the Interoceanic Region Authority (ARI). ARI was created by the Panamanian government in 1993 to oversee and promote investment in the canal region. One of ARI's objectives is to develop tourism opportunities within the Interoceanic region (ARI 1997b). ARI has chosen to emphasize general tourism over what would typically be defined as ecotourism. Currently, ARI's is promoting Law 8, along with specific ecotourism projects, to increase tourism (ARI 1997b, 1998). ARI is promoting at least one ecotourism venture in the town of Gamboa, a former military base. Just 45 minutes away from Panama City, Gamboa lies between the Lake Gatun and Soberenia National Park. ARI is also encouraging nature-tourists, along with business travelers, to stay at the converted Former School of Americas.

In addition to the master plan, the Action Plan for the Development of the Tourism-Conservation-Research Strategic Alliance for the Republic of Panama (TCR) has been adopted by the Institutional de Recursos Naturales Renovables (INRENARE), the President's National Office for Science and Technology, the Smithsonian Tropical Research Institute, and IPT (Ayala 1997b). TCR is a proposal for "heritage" tourism of Panama's natural and cultural resources. It outlines a partnership between STRI's research on Panama's natural resources, tourism, and conservation groups to achieve resource conservation. TCR's main focus is on establishing eco-resorts and then requiring the donation of a portion of each eco-resort's profit to STRI.

Can Ecotourism Provide National-Scale Economic Benefit?

For Panama, the presence of the US in the Canal Watershed has historically provided a strong source of national revenue. As U.S. military bases withdrawal, Panamanians will lose an estimated 6,000 jobs (Rigole and Langlois 1996). According to some estimates, spin-off from US presence in the canal watershed economically benefits 500,000 Panamanians (Rigole and Langlois 1996). Many organizations, such as ARI, INRENARE, and IPT have promoted ecotourism as a panacea for economic and conservation challenges of the Watershed. But can tourism within the Canal Watershed provide a national economic substitute for the US military presence? Tourism can contribute positively to national development, but in many cases, ecotourism has not met economic expectations (Brandon 1996). This is because, by definition, ecotourism is limited economically. If ecotourism occurs on a scale large enough to provide a national economic benefit, the natural and cultural resources it depends upon will be deteriorated by intense visitation (Langholz 1996). As parks become crowded and degraded, and local peoples lose their native culture, the demand for ecotourism lessens, as does the economic benefit. There are few examples of countries where ecotourism provides a major income source, and even fewer that are operating on a sustainable level. Some might argue that Costa Rica's national economy benefits from ecotourism in a sustainable way. However, although ecotourism has surpassed bananas as the country's number one export, many of Costa Rica's protected areas have suffered from severe trail erosion, water pollution, and changes in wildlife behavior (Norris 1994).

Therefore, it is unlikely that ecotourism will provide the scale of economic benefits needed for the Panamanian people. From a national level, ecotourism can provide economic benefit, but only as part of a comprehensive plan. Other tourism initiatives should be examined. Because they can be carried out on a larger scale, general tourism and nature-based tourism have potential to generate a national income supplement. Although tourism and nature-tourism will have some impact on natural and cultural resources, they can be implemented on a scale large enough to generate substantial economic benefits. Because of this, general tourism and nature-tourism can meet some of the goals of ecotourism. For example, a portion of tourism profits can be used to mitigate for impacts on protected areas due to higher visitation. There is a place for true ecotourism projects, especially in community-based ecotourism projects and very fragile ecological areas, both of which can only support small groups of toursits. However, other activities can provide for larger amounts of nature-based tourism and mass tourism. So, the question is: can a combined ecotourism, nature-tourism, and general tourism industry within the canal watershed generate a national-level economic benefit, while also achieving some of ecotourism's goals?

Several things must be considered when debating whether tourism can provide national benefits. First is Panama's ability to attract a large tourist base. Extensive cultural and natural features of the country exist, but these need to be promoted. In addition, there are two other considerations: tourism's ability to generate large amounts of wide-spread income and Panama's capacity to support large numbers of tourists.

The prospect of foreign exchange generation and national revenues are the strongest reasons for developing countries' interest in tourism (Brandon 1996). These development goals are often met. Travel and tourism is the world's largest industry (World Travel and Tourism Council 1995), and projected tourism earnings for 1996 were $3 trillion dollars worldwide (Taylor 1996). Likewise, nature-based tourism can be a major force of economic development, generating foreign exchange, employment, economic diversification, and regional growth (Goldfarb 1989). And the demand for general tourism and nature-based tourism is increasing worldwide, as well as in Panama. In Panama, tourism had a 12% growth rate between 1990 and 1997, and it is currently responsible for 6% of employment (Tribaldos 1997). At the same time, the demand for nature-based tourism has expanded (Brandon 1993). Given the current trends, it is likely that some combination of ecotourism, nature-based tourism, and general tourism will provide a strong economic benefit to Panama.

To make the tourism industry profitable, Panama must be able to support large amounts of tourists. Tourism plummeted during the Noreiga years, but visitation rates have increased since 1992 (Rigole and Langlois 1996). Currently, 16% of all visitors to Panama are tourists (IPT 1993). If ecotourism and nature-based tourism are promoted, overall tourism rates will increase. A major challenge associated with increased visitation is the attainment of capital to provide infrastructure and services. Solutions to keeping investment, and therefore profit, within the country are outlined in the Recommendations section. Another challenge is the prevention of degradation to natural and cultural areas. A hundred visitors a year may have a negligible effect on parks and local peoples, but 1000 or 10,000 certainly will impact resources. It is easier to maintain the ecological integrity of the environment when tourism volume is kept small. The watershed provides numerous tourism attractions, some of which can support large numbers of visitors. However, it is paramount that tourists have limited access to sensitive cultural and natural areas.

To create a truly national benefit, economic profits must be distributed throughout Panama's population. A first step to achieve this is to ensure that most of the investment in tourism and ecotourism projects is from Panamanians. Often, ecotourism projects in developing countries are funded by foreigners (Brandon 1993). Therefore, although some locals benefit from wages, the economic profit leaves the country and does not add to national wealth. So, a tourism strategy should take measures to ensure that projects are owned and operated by Panamanians. But foreign investment is not the only reason money often leaves the country. Often, many profits leave the country in the form of "economic leakages" (Brandon 1993). This is tourism jargon for the money that flows out of a country in order to support tourism. For example, Panamanian ecotourism projects may require oil importation, foreign-owned airlines, and skilled professionals such as hotel managers and travel agents, among other foreign goods and services. When leakages are taken into account, countries have much lower tourism earnings. Studies have found that only 10% of tourism spending remains in Zimbabwe (Lindberg 1991) and that 10-20% of all tourism proceeds are retained in Jamaica (Church and Brandon 1995). Therefore, to keep money within the country, Panama's tourism strategy should include ways to decrease leakage. Lastly, the tourism strategy should strive to spread economic benefit throughout the people living within the canal watershed, instead of concentrating wealth in a few elites.

RECOMMENDATIONS

Planning an Ecotourism Strategy

Although Law 8, Panama's master plan, and TCR are all ambitious, none of them establish a strategy for protecting natural and cultural resources, encouraging community-development, or distributing profit throughout Panama's population. These key objectives must be met for ecotourism success.

Otherwise, if the Panamanian public does not have an interest in preserving natural and cultural resources, forested land will be used for other, more economically beneficial uses (Cope 1996). The following are some recommendations for how these ideas can be implemented.

Thorough understanding of social structure-In order for Panama to reach its objectives of conservation and economic development through ecotourism, a thorough understanding of the social structure of the participating communities is imperative. Cohesive communities with distinct self-identities are perhaps the best suited for ecotourism ventures. It is also important to understand and monitor the existing political stability of the country as can be seen by the Rwandan example. However, at this point there are no political problems that would lead us to believe that an ecotourism plan would be in jeopardy.

Consider carrying capacity-Development cannot be allowed to run rampant. The ecotourism program must be scaled to the carrying capacity of both the natural resource and the community that the program is designed to promote and/or protect. For example, the Embera village in Chagres National Park is an ecotourism project whose success is due in part to the remoteness of their village. Tourism is encouraged, and revenue is generated, by the selling of native arts and crafts. However, due to the limited access and difficult journey that one must embark upon to reach the village, tourism has been kept at a sustainable level. Barro Colorado Island is an example of an ecotourism project designed around a natural resource. Tourists are allowed to visit this protected area but access is limited and controlled by reservations made well in advance. This is done in order to maintain the natural integrity of the island and to prevent an influx of tourists from degrading the forests. Regardless of the scale of the project it is important to ensure that the resource, either natural or cultural, is not overexploited. One way to ensure that resources are not overexploited is a restructuring of zoning ordinances. Large eco-resorts could be limited to areas with existing infrastructure, while protected areas could be reserved for conservation.

Use of existing infrastructure-The lack of capital to build infrastructure is a limiting factor in both ecotourism and tourism. Panama is in the unique position of having hundreds of former military buildings. Using existing Canal Zone infrastructure would allow for Panamanian nationals to keep control of many ecotourism projects. This would allow more money to stay in-country, in the short run because "leakages" would be reduced, and in the long term because projects would be domestically owned. An additional advantage is that fewer forested areas would have to be cleared to build hotels, inns, restaurants, and other tourism facilitates. Converted former military buildings would be excellent choices for nature-tourism, since they are close to the various national parks, as well as other attractions in the canal area. This could be achieved by giving tax breaks or subsidies to projects that use existing infrastructure.

Create incentives for domestic projects–The Panamanian government has already created general incentives for tourism projects. To reduce leakages, these economic incentives should focus on helping domestically run projects. Tourism needs to be promoted, so economic incentives should be kept, but they should be made more favorable for domestic investment. For example, the real estate tax exemption could be kept at the 20 year level for foreign investors, but extended to 30 years for domestic investors. Likewise, in order to compete with wealthy international investors, it is key that domestic investors receive credit, such as low-interest loans, to begin projects.

In addition, incentives can be used to achieve other goals. Tax reductions and subsidies could be given to projects that use Panamanian products and services, such as food, travel agencies, and building supplies. Models from other countries can be helpful. For instance, Columbia has established a legal and organizational framework for supporting independently owned nature reserves (Cardenas 1994, Government of Columbia 1993). Independently owned nature reserves often have strong ties to community-based conservation efforts (Langholz 1996), provide permanent and temporary employment, and preserve previously unprotected land (Langholz 1996).

Incentives for domestic nature tourism projects–The Panamanian government has already created general incentives for tourism projects. To reduce leakages, these economic incentives should focus on helping domestic-run projects. Tourism needs to be promoted, so economic incentives should be kept, but they should be made more favorable for domestic investment. For example, the real estate tax exemption could be kept at the 20 year level for foreign investors, but extended to 30 years for domestic investors.

In addition, economic incentives can be used to achieve other goals. Tax reductions and subsidies could be given to projects that use Panamanian products and services, such as food, travel agencies, and building supplies. Likewise, in order to compete with wealthy international investors, it is key that domestic investors receive credit, such as low-interest loans, to begin projects.

Establish a skilled domestic workforce–To keep the majority of nature tourism profit within Panama, domestic development of projects must be encouraged. But this alone will not ensure that nature tourism projects will benefit the general Panamanian public. To achieve this, a skilled domestic workforce must be created. Travel agents, hotel managers, interpreters, and other skilled positions are better paid than the general service jobs to which domestic laborers are often limited. Because of the international presence within the country, many of Panama's citizens have the advantage of being familiar with foreign customs and languages (ARI 1998). However, more education would increase specific knowledge of tourism service skills. Some initiatives have already been established, such as an agreement with the Swiss School of Hotel Management, which is being developed in the former US military base of Espinar. The school will enroll 350 Latin American students a year in hotel management courses (ARI 1998). Other domestic worker training programs need to be negotiated and/or funded by the government. For example, there is a huge tourist demand for both traditional and modern knowledge about natural and cultural resources. This is a perfect opportunity to involve locals as tour designers and guides.

Monitoring–In order to maintain any successful program it is important to monitor and examine the aspects of the program on a regular basis in order to ensure that the program is still meeting its specific objectives. Additionally, continuing education of those overseeing the project can help to guarantee that specific goals and objectives will be met. If it is found that the program is not meeting goals and objectives it should either be terminated or revised in order to maintain the resources it was designed to protect.

A review of recent literature can help establish some monitoring guidelines. Principles and methods of monitoring are summarized in Wallace 1996. Brandon and Margoluis (1996) underscore the point that monitoring must not only include the collection of data, but also an analysis of the impacts resulting from collected information. For example, protected area managers should record the number of educational programs presented to visitors, but this information should be linked to the economic benefits resulting from the creation of an educational staff, and the environmental degradation that may result from visitation. Monitoring does not have to be limited to only park employees. Several new projects in Guatemala are successfully promoting monitoring by the very ecotourists who create some of the impact (Talbot and Gould 1996). In addition, Cespedes (1996) summarizes the use of ecotourist evaluations in Costa Rica.

It may be useful to assign appraisal of Panama's ecotourism and tourism strategy and programs to an independent group. This group could consist of a committee made of members from local community groups, major agencies (ARI, INRENARE, IPT, etc.), and NGOs (ANCON, STRI, etc.).

Establish a certification system–A certification system to distinguish between different levels of ecotourism could be used in conjunction with committee. For example, project operators who wish to be certified as "ecotourism ventures" must comply with standards based on environmental impacts, financial contribution to resources, and education to visitors. This would help to lessen impacts to and increase economic benefits for natural and cultural resources. Ideally, visitors would chose to use services, such as lodges and tour programs, that comply with the certification program. The Ecotourism Society has already established a ranking system that could be easily adapted to Panama's program (Ecotourism Society 1993). The appraisal committee could be responsible for certification and monitoring of certification requirements. To ensure success, tourists must be made aware of the certification program.

Ensure that a percentage of the profit is reinvested in the natural and cultural resources–The government can use several tools to capture revenue for natural and cultural resources. User fees can be charged to people who use a protected area, but for the resource to benefit, there must be a requirement that a portion of tour profits are returned to the site. This can either be done in two ways. A portion of the user fees generated at a specific location can be returned to that same protected area or facility. Or, a general protected area fund can be established, and a portion of the fund is designated for each protected area. In addition, goods and services used by visitors can be taxed. Food, hotel, and airport taxes can be added to the general protected area fund.

Income for the preservation of natural and cultural resources can also be generated from project operators. Concessions are fees charged to groups of individuals who are licensed to provide services to tourists at specific site (Brandon 1996). For example, services such as food, lodging, transportation, and guide services can be granted concessions. In addition, sales and royalties from activities or products of a tourism site can be collected. Examples include postcards, photographs, and pharmaceutical products.

Involve local communities, the private sector, and the government–A successful ecotourism venture cannot be accomplished without support from all levels. Local involvement is required to balance economic benefits and environmental impacts. The private sector, including universities, NGOs, and even large corporations, can lend expertise and resources to tourism projects. And the government has a crucial role in planning, promoting, implementing, and regulating ecotourism projects. Lastly, partnerships between all three sectors is crucial.

REFERENCES

Ayala, H. 1997a. Panama: the challenge of making history. EcoResorts International, Irvine, California.

Ayala, H. 1997b. Resort Ecotourism: a catalyst for national and regional partnerships. *The Cornell Hotel and Restaurant Administration Quarterly* 38(4): 34-45.

Autoridad de la Region Interoceanica (ARI). 1997a. Memoria 1996-1997. ARI, Panama City.

Autoridad de la Region Interoceanica, 1997b. Panama Facts & Figures. Agency brochure. ARI, Panama City.

Autoridad de la Region Interoceanica. 1998. ARI's website. www.ari-panama.com

Arnold, D. 1995. Surviving human chaos. *WorldView* 8(3):5-10.

Brandon, K. 1993. Basic steps in encouraging local participation in nature tourism projects in ecotourism: a guide for planners and managers. The Ecotourism Society, Alexandria, Virginia.

Brandon, K. 1996. Ecotourism conservation: A review of key issues. The World Bank. Paper No. 33.

Brandon, K. and R. Margoluis. 1996. The bottom line: getting biodiversity conservation back into ecotourism. In: pp. 28-38 Joseph Miller and Elizabeth Malek-Zadeh. The Ecotourism Equation: Measuring the Impacts. Yale School of Forestry and Environmental Studies Bulletin 99.

Cardenas, O.O.A.N. 1994. Red de reservas naturales de la sociedad civeil (unpublished). Cali, Columbia.

Ceballos-Lascurain H. 1993. Ecotourism as a world-wide phenomenon. In: Ecotourism: A guide for planners and managers. K. Lindberg and D.E. Hawkins (eds). The Ecotourism Society, North Bennington Vermont, pp. 12-14.

Cespedes, C. 1996. The use of client evaluations in the ecotourism process: an example from Costa Rica. In: pp. 153-159. J. Miller and E. Malek-Zadeh (eds). The Ecotourism Equation: Measuring the Impacts. Yale School of Forestry and Environmental Studies Bulletin 99.

Church, P. and K. Brandon. 1995. Strategic approaches to stemming the loss of biological diversity. United States Agency for International Development, Center for Development Information and Evaluation, Washington, D.C.

Cope, G. 1996. Nature travel and rainforests. In: J. Miller and E. Malek-Zadeh (eds.), The Ecotourism Equation: Measuring the Impacts. Yale School of Forestry and Environmental Studies Bulletin 99:43-48.

Ecotourism Society. 1993. Ecotourism guidelines for nature tour operators. The Ecotourism Society: North Bennington, Vermont.

Ecotourism Society. 1998. Ecotourism Society's web page. www.ecotourism.org

Endara, M. 1997. Taking conservation of Panama's natural resources into the XXI century. EcoResorts International. Irving, California.

Fennel, D. and P. Eagles. 1990. Ecotourism in Costa Rica: A conceptual framework. *Journal of Park and Recreation Administration* 8(1):23-34.

Foster, J.W. 1992. Economic considerations in mountain gorilla conservation. In: pp. 87-91. Proceedings Joint Meeting AAZV/AAWV.

Goldfarb, G. 1989. International Ecotourism: A strategy for conservation and development. The Osborn Center for Economic Development, World Wildlife Fund-Conservation Foundation, Washington, D.C.

Government of Columbia. 1993. Ley del Ministerio del Medio Ambiente, articulos 108-111.

Government of Panama. 1992. By the resolution of the Cabinet No. 46 of February 19, 1992.

Government of Panama. 1996. Legislativa, Ley No. 8. Gaceta Oficial, Year XCI, No. 22, 558: 1-25.

Instituto Panameno de Turismo (IPT) and Organization of American States (OAS) 1993. Plan maestro de desarrollo turistico de Panama: 1993-2002, Panama City.

Haxton, M.L. 1995. Native peoples of Panama. CIA Factbook, Washington, D.C.

Langholz, J. 1996. Ecotourism impacts on independently owned nature reserves in Latin America and Sub-Saharan Africa. In: pp. 153-159. J. Miller and E. Malek-

Zadeh (eds). The Ecotourism Equation: Measuring the Impacts. Yale School of Forestry and Environmental Studies Bulletin 99.

Lindberg, K. 1991. Policies for maximizing nature tourism's ecological and economic benefits. The Ecotourism Society, Washington, D.C.

McNeil, J. 1996. Costa Rica: The Rough Guide. Penguin Books, London.

Meadows, D. 1995. Beyond Shamans, Toucans and Tourist: Local Participation in Ecotourism in Ecuador and Costa Rica. Proceedings of the 1995 meeting of the Latin American Studies Association.

Ministry of Labor and Social Welfare, 1996. Employment and Training System Project submitted by the Government of Panama (MLSW) to the United States Agency for International Development.

Norris, R. 1992. It's green. It's trendy. Can ecotourism save natural areas? *National Parks* 66(1-2): 30-34.

Norris, R. 1994. Ecotourism in the national parks of Latin America. *National Parks* 68: 33-37.

Reed, S. 1995. Survivors in the mist. *People* March 6:40-45.

Rigole, M. and C. Langlois. 1996. Panama. Ulysses Travel Publications. Montreal.

Sproule, K. 1996. Community-Based Ecotourism Development: Identifying Partners in the Process. In: pp. 153-159. J. Miller and E. Malek-Zadeh (eds). The Ecotourism Equation: Measuring the Impacts. Yale School of Forestry and Environmental Studies Bulletin 99.

Talbot, B. and K. Gould. 1996. Emerging participatory monitoring and evaluation programs in two ecotourism projects in Peten, Guatemala. In: pp. 95-107. J. Miller and E. Malek-Zadeh (eds). The Ecotourism Equation: Measuring the Impacts. Yale School of Forestry and Environmental Studies Bulletin 99.

Taylor, M.R. 1996. Go, but go softly. *Wildlife Conservation* March-April: 12-19.

Tribaldos, C. 1997. Tourism champions conservation of a world's crossroads heritage. EcoResorts International. Irving, California.

Wells, M. and K. Brandon. 1992. People and Parks: Linking Protected Area Management with Local Communities. The World Bank.

Wheat, S. 1994. Taming tourism. *Geographical* April:16-19.

Wallace, G. 1996. Toward a principled evaluation of ecotourism ventures. In: pp. 119-140. J. Miller and E. Malek-Zadeh (eds). The Ecotourism Equation: Measuring the Impacts. Yale School of Forestry and Environmental Studies Bulletin 99.

World Tourism Organization (WTO). 1996. Central American presidents look to WTO. *WTO News*, 2:16.

World Travel and Tourism Council (WTTC). 1995. Travel and tourism's economic perspective. A special report from WTTC.

Ziffer, K. 1995. Ecotourism: Paradise gained or lost. www.oneworld.org/panos/ panos_ eco2.html.

Mutual Incomprehension
or Selective Inattention?
Creating Capacity for Public Participation
in Natural Resource Management
in Panama

Benjamin Gardner

SUMMARY. The Panama Canal Watershed and the Inter-Oceanic Region support a large percentage of Panama's economic productivity. Much of the management of the area is driven by singular perspective goals, leading to conflicting strategies and actions that benefit special interests over the common interest. This paper discusses how myth or normative views of the world often drive narrow problem definitions and preclude broad participation. It describes how trends and conditions in the conservation movement in Panama have shaped perspec-

Benjamin Gardner received a Master of Environmental Studies degree at the Yale School of Forestry and Environmental Studies, New Haven, CT 06511. Currently he is Consultant for JICA (Japanese International Aid) in Panama.

The author would like to thank the staff of INRENARE/ANAM for hosting his visit and sharing their insights while in Panama. The author would also like to thank all of the people and organizations that took time out of their schedules to meet and talk with him. These include; STRI; PCC; ARI; Fundacion Natura; ANCON; Professors from the University of Panama; Mario Bernal; Takayuki Hagiwara; Mirei Endara; the residents of Agua Buena, La Bonga and San Juan de Pequeni; and others. He would also like to thank Dr. Clark, Dr. Ashton and Jennifer O'Hara for their support and encouragement in carrying out this project.

[Haworth co-indexing entry note]: "Mutual Incomprehension or Selective Inattention? Creating Capacity for Public Participation in Natural Resource Management in Panama." Gardner, Benjamin. Co-published simultaneously in *Journal of Sustainable Forestry* (Food Products Press, an imprint of The Haworth Press, Inc.) Vol. 8, No. 3/4, 1999, pp. 127-145; and: *Protecting Watershed Areas: Case of the Panama Canal* (ed: Mark S. Ashton, Jennifer L. O'Hara, and Robert D. Hauff) Food Products Press, an imprint of The Haworth Press, Inc., 1999, pp. 127-145. Single or multiple copies of this article are available for a fee from The Haworth Document Delivery Service [1-800-342-9678, 9:00 a.m. - 5:00 p.m. (EST). E-mail address: getinfo@haworthpressinc.com].

tives, and then poses several alternatives to achieve a more inclusive natural resource management decision process in the region. *[Article copies available for a fee from The Haworth Document Delivery Service: 1-800-342-9678. E-mail address: getinfo@haworthpressinc.com <Website: http://www.haworthpressinc.com>]*

KEYWORDS. Natural resource management and policy, land use, participation, myth, conservation and development, Panama Canal Watershed

INTRODUCTION

The Panama Canal plays an essential role in Panama's economy. The Panama Canal Watershed (PCW) and the Inter-Oceanic Region, the areas surrounding the Canal, support approximately 80% of the country's GDP and are home to over 50% of the country's population. The Canal itself contributes between 6% and 10% to Panama's GDP and employs 12,000 people (EIU 1995, Mortimer 1987). While it is unclear who the coordinating management authority within the Watershed is (see Khan, this issue), several agencies have initiated strategies to address issues of land-use and development in the region.

While some of the current strategies being employed address specific concerns of land conservation and economic development, there is a lack of coordination as each intervention is being guided by a particular organization's perspective. In other words, how an organization views the world, which is heavily influenced by its assumptions or myths, leads the individuals who work for it to frame a particular problem in a manner that will best serve their mission or goal. While in many cases these organizations are well-suited to define problems and select alternatives, they are often driven by self-serving goals that do not ultimately account for the perspectives of Watershed residents and may fail to address the common interest for the area's long-term sustainable development. In this paper I will elaborate on this policy problem, outline trends and conditions that have helped shape the current situation, and propose alternatives and recommendations. While I try to encompass the overall resource management process in the Panama Canal Watershed and the Inter-Oceanic Region in my analysis, my recommendations will focus on the past role of the Institute of Renewable Natural Resources (INRENARE) and its new replacement the National Environmental Authority (ANAM). I take this approach primarily because INRENARE has played a central role in managing the natural resources of the Panama Canal Watershed and it appears that ANAM will play a pivotal role well into the future.

THE PROBLEM

After reviewing the mission and goal statements of several organizations[1] working in the Panama Canal Watershed and the Inter-Oceanic Region, the overall management goal for the area can be summarized as follows: to ensure economic development and promote human dignity, while managing the natural resources for both short-term utilization and long-term sustainability. However, the current emphasis on economic development and short-term utilization of natural resources, over human dignity and long-term sustainability, jeopardizes Panama's ability to meet its goals for the Canal Watershed and the Inter-Oceanic Region.

The problem is that the current resource management process in Panama is driven by problem definitions that are derived from singular and narrow perspectives. These problem definitions are often in direct conflict with each other, yet are selected with little regard to others' goals or values. For example, the recent creation of a land use law (#21-1997) for the Panama Canal Watershed, is based on a definition of appropriate land use that was defined by policy makers using a certain set of criteria. These criteria resulted in an analysis that stated that a large percentage of current land use in the region is inappropriate and therefore must be changed. However, many residents of the PCW do not view their land use as inappropriate and therefore see little benefit in changing their practices. The estimation of problems within such a context often excludes multiple perspectives and fails to achieve the necessary support to meet management goals and objectives. I believe that a more inclusive decision process, while necessitating compromise, will lead to more balanced and sustainable land-use management in the area.

METHODS

My analysis is based on research carried out as part of a semester-long graduate seminar titled "Forest Conservation for Productivity and Diversity." As part of this course, I spent ten days in Panama meeting with natural resource officials and field staff, community members, nongovernmental organization staff, and independent researchers. My methods included informal interviews carried out in a group setting with sixteen students from diverse backgrounds in natural resources management, a literature review, as

---

1. Documents consulted include: the Panamanian Constitution; the mission statements of the Inter-Oceanic Authority, the Smithsonian Tropical Resources Institute and the National Association for the Conservation of Nature; the mission statement and objectives of the Institute of Natural Renewable Resources; and the organic acts for the Panama Canal Commission and the Panama Canal Authority.

well as analysis of several policy documents. I have used a policy framework to help organize and present my analysis. Although much of the information collected in the field is highly impressionistic, as our time and exposure were quite limited, I hope my analysis will provide insights and be useful to those interested in natural resource management in Panama.

MYTHS AND PERSPECTIVES

While there are many participants involved in land-use decision making in Panama, I will highlight only a few in order to illustrate how the social process affects resource management decisions (for more information on participants see Whitney, Williams and Maxwell, and Khan, this issue). Table 1 describes some of the relevant participants to the management process and the prevailing myths to which they subscribe. These myths are my interpretation and are clearly generalizations, however, I believe that they do speak to the perspectives or normative views of various participants.

A myth is a belief or perspective that is so widely or fully held, that it is often imperceptible to those who believe it (Lasswell 1971).

> Myth is a term of art designed to refer to all relatively stable and coherent patterns of perspective. (The implication is not that these perspectives are necessarily true or false in the empirical sense.) Every individual or group develops a distinctive myth . . . By the formula is meant the prescriptive norms of conduct that are adhered to on pain of deprivational sanction. The miranda are the relatively concrete and expressive elements . . . (Lasswell 1971)

Myth is often based on narratives that are told and created over a period of time. ". . . Development narratives [are] stories about the world that frame problems in particular ways and in turn justify particular solutions" (Leach et al. 1997). Myth is often based on a real perception or at least a perception that was credible within a particular historical context. Individuals and organizations often hold a set of myths that help to make up their worldview.

While myth may or may not be based in fact, it is an influential factor in the way people and organizations view the world. It is through myth that we can begin to understand diverse perspectives and the decisions that they lead to. None of the myths or worldviews in Table 1 is mutually exclusive or total. They are in fact only partially unique from each other. Within these myths there is overlap, and thus, the opportunity for greater partnerships in order to achieve mutually beneficial goals. However, many myths contradict others' myths, and thus cause barriers to participation and dialogue. For example, the myth of the pristine Watershed remains one of the driving

TABLE 1. Stakeholders within the Panama Canal Watershed and their associated prevailing myths.

Stakeholder	Myth	Description
Popular environmental myths	*The myth of the pristine watershed.*	Before 1904 the Watershed was "natural," people then "invaded" and "degraded" the environment. Since people are the main threat to the Canal, institutions must protect the Canal from people.
Popular environmental myths	*The Canal will dry up! "deforestation" has led to reduced rainfall. Trees bring rain.*	While these perceptions are not based on credible scientific data for the area (Wadsworth 1978), they are widely accepted and even promoted informally. While trees may have some macro-climactic effects in certain regions such as the Amazon or Congo Basin, they do not in a small country like Panama where ocean weather systems are the main factors on precipitation (Intercarib S.A. and Nathan Associates, Inc. 1996; Pereira 1989).
Institute for Natural Renewable Resources (INRENARE) National Environmental Authority (ANAM)	*If left to their own devices, people will naturally harm the environment.*	People do not have the scientific knowledge to carry out responsible and modern resource management or conservation. It is the responsibility of trained professionals to work with local people to manage the country's natural resources.
Panama Canal Commission (PCC)	*We currently have enough water to run the Canal, if we need more water we will create or divert a new source such as a reservoir or an additional river.*	They are concerned with passing ships through the Canal and current land use conditions in the Watershed are not seen as a significant harm to efficient Canal operations.
Rural Resource Users	*The government will do what they want, they don't care about us.*	"We want loans to raise chickens and pigs, and to reforest to control the grass" (La Bonga community member 1998). Years of conflicting policies and authoritarian rule have left many people disillusioned with the government.
Inter-Oceanic Regional Authority (ARI)	*Economic development is the most important goal.*	ARI's mission is to develop the Inter-Oceanic Region and then dissolve. Their primary concern is generating employment. ARI hopes to lay the "building blocks toward the making of a new mini-Singapore" (Mejia 1998).
Fundacion Natura (The Nature Conservancy, The United States Agency for International Development and the Panamanian Government)	*Trees are good! If people plant trees, then people are good. Let's encourage people to be good!*	The problem has been defined and the answer is to plant trees.
Ministry of Agriculture (MIDA)	*You can't eat trees!*	Agricultural development is their mission and way of life.
Panama City Homeowners	*Trees are good neighbors!*	Some want to build second homes in the Watershed, and they often have the political and economic power to do so.
Nongovernmental Organizations (NGOs)	*We speak for the trees and for the people!*	NGOs believe that certain perspectives are not being accounted for, and they want to bring them to the discussion.

conceptions behind several current management decisions in the PCW. This myth is conveyed by the following statement, but is repeated in many forms by various participants in the Watershed.

It is commonly stated that the first large migration of people to the Watershed occurred during the construction of the railroad in the mid-1800's followed by the Canal in the early 1900's (Heckadon, 1993, 1998). "This was the first invasion of people. . . . The area was declared a national park and that is why it is still intact . . . the upper Chagres Natural Park has never been cut" (Alvarado 1998). While this most recent migration was the first large influx of human population in centuries, historical data shows the high probability of large indigenous populations living in the area thousands of years ago (Illueca 1985, Posey and Balee 1989). The current narrative, while based on real events of the past century, denies that the Watershed has been managed by humans for thousands of years. The myth of the pristine Watershed creates the impression that there is some pristine status to which the area could or should be restored. While seemingly harmless to many conservationists, this myth can lead to poor management decisions with "unintended consequences," and waste labor, capital, and natural resources that could be more useful elsewhere (Current 1994). While many organizations see rural land users as poor land managers in need of education and modernization, this perception is contrasted with the common use of their image as a marketing tool for national parks and Watershed management. It is not unusual for an institution to take away the rights of campesinos or indigenous people "for the good of the country," while simultaneously invoking their image as "'natural" people or somehow closer and more in tune with nature.

DECISION PROCESS

Pressure from both international donors and internal organizations has led to an increased discussion of participation in conservation and development efforts in Panama. In fact, much of the current discourse concerning the conservation of natural resources in Panama includes a greater integration of public values and demands in the decision making process (Cortez-Hinestrosa 1998). However, many of the agencies responsible for implementing these programs in Panama do not currently have mechanisms to incorporate diverse perspectives and goals into their decision-making processes. I believe that the overall management goal (see Introduction) of the area cannot be achieved without substantive public participation, and that the reversion of the Canal and lands in the Inter-Oceanic Region present an incredible opportunity for a new system of dialogue on natural resource management issues.

All management or policy processes are made up of a "multi-step," interconnected decision process. Whether explicit or not, these decision activities guide management actions and determine if they will be successful (Clark 1995, Lasswell 1971). One of the key elements in decision making is defining goals and objectives. Goals are policy statements by organizations reflecting their perspectives and how they hope to affect change. A management process is often driven by an organization's goals.

In clarifying the management goal, it must be asked: whose goal is it and who is involved in clarifying it? Can the common interest be achieved if only one perspective is being taken into account? Myths drive assumptions and the estimation of problems, which often alienate certain groups from participating. A "healthy" decision process encourages open dialogue and provides checks and balances by reflecting multiple values and goals.

PARTICIPATION

Participation can be defined as the inclusion of politically relevant actors in the decision process (Lasswell 1971). This seems simple enough, but the difficult questions follow. Who decides relevance, and why should those in power want to share it? The ability of groups or individuals to deprive or indulge those in power of important values is often the method by which participation is increased (Reisman 1995). These value indulgences and deprivations can take many forms depending on the context of the situation (Clark 1995, Reisman 1995). It is important to distinguish that participation does not equal representation. Ultimately, what is important is if a participant's goals and values are considered in management decisions, not whether they are present at a meeting.

The goal of participation is to strengthen the decision process by broadening the goals and perspectives considered. Increased participation should promote human dignity and social justice and in turn increase the chances of meeting long-term technical management objectives. It is hoped that increased participation will begin to break down artificial barriers between categories such as the "state" and "communities," and increase trust leading to a stronger and more stable system that will meet the needs of multiple actors. Evaluations of past development and conservation efforts have shown that no one technical solution can provide "the answer," and that the best way to ensure accountability of those in power is to increase dialogue on land use issues in general (Wells and Brandon 1992).

There are many interpretations of participation, from consultation and education to having a voice in management decisions. It appears that participation in Panama is being interpreted as the former rather than the latter. I contend that to meet the ultimate management objectives for the Watershed and Inter-Oceanic Region, stakeholders will need to participate at a more sophisticated level in the decision process. The belief by managing agencies that participation will dilute their goals and weaken their ability to enforce policy is one of the main constraints to higher level participation. While this type of participation may appear more threatening to resource managers, it could ultimately lead to a more stable system of management in which conflicts can be identified and addressed early on.

Although participation will most likely necessitate a compromise, it may offer the best way for an organization like INRENARE or ANAM to meet their objectives. Participation is not a guaranteed solution; it is an attempt to help address a complex management problem, which has yet to find an effective and sustainable management regime (Agrawal 1997). While most discussions of development and resource management focus on institutions and policy documents, I argue that management is a process that is ultimately negotiated on the ground between local resource users and field managers, not by policy makers in offices in Panama or other cities.

A commitment to collaboration and active participation by local people means a commitment to initiating a new decision process. A decision process that engages local people should seek to incorporate their views from the beginning of a management cycle. This includes their participation in initiating the project, including the selection of relevant stakeholders: estimating the problem; helping to define the management goal; selecting various management options; helping to implement projects; evaluating the management actions; and helping to determine if a certain management practice should continue, be refined, or be terminated outright (Clark 1995).

I am not arguing that the government agencies charged with management should give over their role to community members. In fact, community members often have a parochial, special interest view of a given problem as well. Participation does not equal local control. The goal of a participatory process strives to be an inclusive process in which stakeholders who have various values, expectations and demands help shape and refine management decisions that are aimed at serving the greatest good for the greatest number of people for the greatest period of time (Pinchot 1947).

ANALYSES OF THE PROBLEM–
TRENDS AND CONDITIONS

Landscape and Population Dynamics:
A Broader Perspective

For thousands of years people in Panama have cleared forest for swidden agriculture. When the first Europeans arrived in the 1500's, they encountered a landscape that was a mosaic of forests, fields and fallows (Linares and Ranere 1980, Illueca 1985). European expansion in the area led to dramatic changes in both the human population and the physical landscape. Within a fifty year period after the Europeans arrived in large numbers, there was a 90% reduction of the indigenous population, due primarily to warfare and disease (Illueca 1985). An ensuing period of forest regeneration occurred, due in part to the artificially low population density (Heckadon and Gonzalez 1985, Posey and Balee 1989).

The building of the Panama Canal from 1904-1914, led to high levels of immigration to Panama and renewed forest clearing (Heckadon 1993, Gradwohl and Greenberg 1988). With significant state promotion of forest clearance and subsidies for cattle ranching, these trends continued throughout much of this century. From the 1950's through the 1970's the government helped move settlers into forested regions so that they could make the land productive (Heckadon 1998). The promotion of forest clearance and agrarian expansion was in large part driven by the desire to lay claim to the land for the overall development of the state. This was particularly important in the early half of the century, when Panama's sovereignty could have been challenged (Zimbalist and Weeks 1991).

The effect of this trend has been the creation of the myth that people are bad for the environment and will naturally degrade it if not controlled. It has also transformed land tenure and migration patterns throughout the country, and led to the development of markets for certain agricultural products, particularly beef (Heckadon 1993, Heckadon and Gonzalez 1985).

The Need for Better Management:
The Creation of INRENARE and ANAM

The growing awareness of environmental issues by the Panamanian government and the international community in the 1980's led to the recognition for the need of a body separate from the Ministry of Agriculture to coordinate natural resource management efforts in the country. INRENARE was established in 1986. Before this time natural resources were managed by RENARE which was a division of the Ministry of Agriculture. While seemingly a positive decision to separate INRENARE from the Ministry of Agriculture (MIDA), which often promoted programs in direct opposition to conservation efforts, the move left INRENARE with a lack of power and influence as compared to other ministries such as MIDA.

In 1998 Law #41 created the National Environmental Authority (ANAM) which has replaced INRENARE as the government agency responsible for the environment. The question remains does the transformation of this Institute into the newly-formed Authority demonstrate the Government of Panama's improved commitment to natural resource management? It is hoped that this

Authority will have more power and support than the former INRENARE did as an institute. Since ANAM has only recently come into existence in 1998, my analysis of past trends and conditions will focus on INRENARE an organization which has had a considerably longer history managing natural resources in the Watershed area. My recommendations for future management, however, will be designed for ANAM.

The overall management goal of INRENARE as it regards natural resources was to define, plan, organize, coordinate, regulate, and promote the policies and activities of conservation, development, and utilization of the country's renewable natural resources (INRENARE 1997). Natural resource management responsibilities are divided between five divisions. These are forestry, watersheds, protected areas including flora and fauna, environmental impact assessments, and environmental education (INRENARE 1997). These divisions are often at odds with one another, and it can be difficult for the Authority to synchronize programs (for more information see Maxwell and Williams, this issue).

"During the 1980's, [Central American] government natural resource agencies in general were weakened to critical levels by internal and external economic policies, 'structural adjustment' and government size reduction, hindering their institutional growth at a time when many more protected areas were being declared" (McNeely et al. 1994). INRENARE, like many natural resource agencies the world over, has been unable to meet its own goals due, in part, to the lack of financial and technical resources. It will therefore be important for the newly created ANAM to plan according to projected resources and not create management strategies that are unrealistic.

For example, in Chagres National Park there are approximately 4000 residents who are legally residing within park boundaries (Figure 1). While their rights to live in the park are recognized, their permitted land-use activities have been typically decided by INRENARE. While these people may be seen as a threat to the traditional park model, they may actually offer an opportunity to establish a more balanced and sustainable management regime. For example, the presence of people in the park is evidence that INRENARE was taking a more participatory approach than those of a more traditional model where exclusion is the rule.

> In the beginning our focus was biological, and to prevent other people from encroaching. Then we had a different reality: a park with people, agriculture and cattle. We had to work with these people and stop new people from coming in. We have tried to incorporate community organizations into management of both buffer zones and within parks. (Cortez-Hinestrosa 1998)

FIGURE 1. The Embera people of the La Bonga community (population 288) legally reside within the boundaries of Chagres National Park, Panama. Photo credit: Jennifer L. O'Hara.

The effect of this trend is a burgeoning interest in working closely with communities by some of the INRENARE staff, while a continued faith in a top-down management approach remains strong in other staff members. Despite internal efforts to reorganize, the Institute appeared relatively weak in the eyes of the government, international community, and local land owners.

Social Stratification and Political Participation

Most of the institutions involved in natural resource management in the Watershed have not substantively included the public in management decisions. This behavior was in part shaped by the US based conservation movement, as well as by Panama's socio-political situation. "By the early 1970's Panama had one of the most unequal income distribution[s] in Latin America: the highest 20 percent of households earned 61.8 percent of household income, and the lowest 20 percent earned 2.0 percent" (Zimbalist and Weeks 1991). Today, Panama still maintains one of the largest wealth gaps in Latin America (Heckadon 1998). This leads to a highly stratified society in which policy decisions are most responsive to values and demands articulated by the wealthy and politically powerful classes.

Panama's political history, while considerably different from many of its neighbors, highlights the lack of popular participation by peasant classes. Early independent Panama saw the emergence of a wealthy merchant class as opposed to the powerful oligarchies or landed elite's in neighboring countries. Recent history has been heavily influenced by military control of the political process. Power in Panama is highly concentrated in elite family networks which cross through prominent sectors of the economy and politics (Zimbalist and Weeks 1991). This has left a decision process in which non-elite participants do not have the power to demand that their values are considered in management decisions, and thus there is little challenge to the status quo.

The effect of this trend is a lack of dialogue concerning natural resource management issues in Panama. Local participation in this dialogue is fragmented and generally weak.

The Reversion of the Panama Canal and Adjacent Lands

In 1977 the Torrijos-Carter agreement was signed, stating that the Panama Canal and the adjacent lands within the ten mile "Canal Zone" would be returned to Panama on December 31, 1999. It also called for the removal of the US armed forces by that same date. The effect of this is a one-time massive reconfiguration of the distribution of wealth and power in the area.

The Spread of Gringo Grass

Despite the fact that many groups' or organizations' myths contradict one another, there remains at least one myth held in common by most Panamanians in the Watershed. This is the story of the introduction and subsequent spread of *Saccharum spontaneum* or gringo grass and its effect on the landscape. *S. spontaneum* is believed to have evolved in southern Asia (Purseglove 1972). Although the origins of its introduction are unclear (see Hammond, this issue), many Panamanians believe that gringo grass, as it is commonly known, was introduced by the United States to control bank erosion along the Canal.

The grass has grown quickly and easily in particular areas in the Watershed, and many residents are deprived of some value as a result of it. This species can colonize areas that have been cleared and are not under current forest or crop cover. While the grass can be effective at controlling erosion on steep slopes, it has spread to otherwise productive land. Once established the grass is extremely difficult to remove, substantial human labor is required to cut the grass and then plant crops and trees to shade out the new grass shoots (Personal communication, Laval 1998).

The effect of this trend has been the reduction in both productivity and conservation potential for many areas throughout the Watershed. This has also led to strong laws prohibiting the cutting of trees after five years of growth. This law, in turn, has severe implications for sweeping land use changes that may be detrimental to currently sustainable land use practices. Another effect of this trend is the apparent willingness of many Panamanians, including local resource users and state agencies to compromise their goals, in order to overcome this invasive species.

PROJECTIONS BASED ON CURRENT TRENDS

Based on these trends and effects, projections can be made to better understand how current conditions will create new problems and affect future management. Given these trends in natural resource management in Panama, projection's point to increasing difficulty to achieve balanced conservation and development goals.

(1) While many organizations in the Watershed have adopted the language of participation, they often use goal substitution to subvert high levels of participation. Education efforts appear to be the focus of much outreach in the area. While education is vital to any park management program, it can also be interpreted as trying to teach people to think like the managing institution rather than to allow local people to bring their own perspectives and values to management or policy discussions.

> In the past, creating national parks meant displacing rural inhabitants from some areas, moving them to other places or, as the best option, limiting their right to the traditional use of the resources. . . . In the 1970's and 1980's environmental education became a vehicle to create awareness in people living around the conservation units in order to obtain their support for projects or, at minimum, so that they would not oppose them. (McNeely et al. 1994)

It appears that this type of effort may repeat itself. For instance, telling people not to burn the forest because it may damage habitat for flora and fauna may not be interpreted in the same way by people who have made their living by utilizing these "riches" of the forest for production.

(2) Although the conversion of INRENARE to an authority (ANAM) is meant to bring with it additional responsibilities, it is unclear what this new status will mean. It appears unlikely that the organization will gain significant political power or financial resources for natural resource management in the future. Given funding trends it also appears that ANAM will increasingly rely on outside organizations to meet its own management goals.

(3) While the political history of Panama has been erratic and often author-

itarian, current trends point to the promise of increased democratization in the country. Much effort will be needed to locate, support and develop institutions through which popular participation can be increased. It is important that institutions working in the Panama Canal Watershed seize this opportunity to encourage and join with these efforts. The reversion of the Panama Canal creates a once in a lifetime opportunity for Panama to make a serious statement on how development activities will be carried out. The decisions made over the next several years will in large part determine the future opportunities and options for the sustainable development of the area.

ALTERNATIVES AND EVALUATION

I will propose five alternatives for management in the Panama Canal Watershed. The first alternative is to do nothing and continue on the current management track. As stated earlier, this will most likely result in a singular perspective management process and the failure to even minimally meet the overall management objective. The second alternative is to implement a participatory decision making process or management strategy across the board. This will most likely lead to resistance within organizations by those who do not believe in the approach or feel threatened by it. This approach may also lead to a half-hearted effort and increased hostilities between the public and management institutions. Also, high levels of participation may not be appropriate for all situations, and could jeopardize some currently effective management systems.

The third alternative is to engage in participatory management on a prototype or pilot project basis (Clark 1995). This alternative is the least risky and will provide the greatest opportunity for "practice-based" learning and for the promotion of institutions such as ANAM and their goals (Clark 1995, Clark and Brunner 1996). This option presents a high potential to create successful models and a way to slowly build internal support and gain external attention and possibly funding. It will allow management institutions to recognize if a truly participatory process can help them best meet their management objectives in certain situations.

The fourth alternative is to use gringo grass to create incentives for collaboration. This may also necessitate changing laws that ultimately discourage conservation (see Hauff, this issue) and developing new incentives for collaboration to achieve multiple goals including productivity and diversity. The eradication of gringo grass is a particularly useful mechanism by which to encourage participation. This alternative could help meet simultaneous objectives leading to more trust and new understandings between participants. The fifth alternative is to focus implementation of each project on increased communication with local institutions and the sharing of information between organizations.

This alternative could provide groups who are often at odds, a common task, resulting in higher levels of trust, and leading to a stronger decision process.

RECOMMENDATIONS

My recommendations are to implement prototypes using a combination of the final three alternatives. The following recommendations describe important elements in carrying out successful prototypes of participatory management. These recommendations focus on the role of the new authority, ANAM, but apply to any organization taking an active role in natural resource management in the Panama Canal Watershed.

Implementing Prototypes

ANAM should choose three to four sites in which to implement prototype management efforts. This would be a low risk alternative with several possible benefits. The prototypes would allow ANAM to experiment with different institutional arrangements in terms of working with NGOs, the private sector or community-based organizations. If promoted well there is a good chance that ANAM could capitalize on funding sources that are currently supporting these types of programs.

The prototype project sites should be carefully selected and will ideally include one or more of the following elements: strong community institutions; relatively good relations with ANAM, particularly with an individual who could be identified as a facilitator (Wells and Brandon 1992); and at least one site should include the presence of gringo grass. This final criteria, while not conventional, is a possible point of leverage which many stakeholders can use to create an open dialogue and collaborative management. As I tried to demonstrate earlier, gringo grass can act as an equalizer to set a level playing field for collaboration.

Toward a Better Decision Process

It is important to be aware of all of the decision functions and secure participation at each level. It is essential to collaborate in the beginning steps of information gathering, problem definition and goal clarification. While it is never possible to incorporate all stakeholders, every effort should be made to share the information collected with other parties and engage in constructive debate. This type of process does not mean that local special interest should take precedence over other special interests. There is, however, an explicit focus on giving a voice to those groups who have been traditionally ignored in the decision making process.

Starting with the prototype or pilot projects will enable ANAM to make explicit goal statements in regard to participation. Projects that have included these have achieved a much higher level of success (Wells and Brandon 1992). If goals explicitly refer to desired outcomes, they can compromise more process-oriented activities that may be necessary to achieve those goals. It is important that management goals are explicit as they concern participation.

> Projects with a beneficiary orientation generally set their goals in terms of changes in readily measurable indexes, such as income levels, farm productivity, [etc . . .] . . . Projects with a participatory orientation . . . seek to achieve goals similar to those of beneficiary projects; however, they are oriented more toward establishing a process leading to change that can be sustained after the project ends. (Wells and Brandon 1992)

Projects that have been recognized as successfully adopting a participatory approach have "started with a clearly stated goal of eliciting local participation, and commitment to a process of participation was clearly reflected in the activity choices" (Wells and Brandon 1992).

Working with Local Institutions

Much of the recent research on conservation and development has pointed to the need for substantive participation by local participants. One of the best ways in which to do this is through local institutions. Local institutions are not necessarily the structured organizations that the word "institution" invokes. Leach et al. (1997) describe institutions as, "regularized patterns of behavior between individuals and groups in society." It is up to outside agencies that want to work with local participants to find the appropriate institutional structure to facilitate participation and accountability at the local level (Leach et al. 1997, Agrawal 1997, Sivaramakrishnan, in press). "Local institutions can act as a focus of mobilization among local people and as a link between local people and external organizations, whether governmental or non-governmental organizations" (Wells and Brandon 1992). ANAM must be flexible with the type of institution it is willing to work with, as communities should not be expected to conform to one blueprint model.

One of the common pitfalls of working with local participant groups is the idea that a given community is a homogenous category filled with like-minded people seeking the same goal. Communities are not static, they are dynamic and internally differentiated (Agrawal 1997, Sivaramakrishnan, in press). "To date, a poor understanding of such dynamic institutional arrangements has impeded practical efforts in community-based sustainable development" (Leach et al. 1997). Most of the current efforts to incorporate local participants in sustainable land-use or conservation initiatives in the Watershed are facilitated through grants from Fundacion Natura. While this model can be successful, the project's effectiveness is currently limited by its approach. Their current approach of funding projects for only two years limits the type of organizations that they can support. They are also having a difficult time finding institutions which they see fit to fund (Hanily 1998). Fundacion Natura would be well served to spend more time working in communities and funding projects that support the growth and development of local institutions.

Building Trust Through Mutual Respect and Increased Communication

Ultimately, one of the most important aspects of a participatory process is respect for the participants. If there is not respect between participants, it will be difficult to cultivate the necessary trust to carry out an inclusive decision making process. The control of information is often used as a management tool to deprive one group of the ability to participate. Organizations such as ANAM should attempt to break down these barriers with local participants as well as other organizations. Field staff who are in daily contact with communities should be given more of a role as information facilitators.

One of the ways to facilitate this new emphasis on communication is to carry out a more detailed study of institutional perspectives and myths. This study would ideally identify specific strategies and areas for collaboration.

CONCLUSION

While this analysis has focused on problems with the natural resource decision process in the PCW, there is much to be optimistic about. Panama is in a unique position to capitalize on its strong service based economy and the relatively large number of well-trained and committed people working toward a better decision process in the region. There is a wealth of knowledge from local land-users, managers and scientists that can contribute to this process. Also, the soon to be drafted National Environmental Strategy presents an excellent opportunity to integrate many diverse viewpoints, as they are charged with establishing goals for the country in a participatory planning process. It is important, however, that this effort recognizes that changing the decision process is a long-term commitment and cannot be fulfilled in a limited period of time. I believe that a majority of the stakeholders concerned with the management of the Panama Canal Watershed can achieve their goals for the area, if they are willing to account for each others' interests and perspectives.

REFERENCES

Agrawal, A. 1997. Community in Conservation: Beyond Enchantment and Disenchantment. Paper prepared for the Conservation and Development Forum, University of Florida, Gainesville, FL.

Alvarado, L. 1998. Panama Canal Commission. Presentation to Yale students on March 16. Gamboa, Panama.

Clay, J.W. 1988. Indigenous Peoples and Tropical Forests: Models of Land Use and Management from Latin America. Cultural Survival, Inc., Cambridge, MA.

Clark, T.W. 1995. A Pragmatic Problem-Solving Method. Draft 8/125/95, Yale NRCC.

Clark, T.W. and R. Brunner. 1996. Making Partnerships Work in Endangered Species Conservation: An Introduction to the Decision Process. *Endangered Species Update*: 13:9.

Cortez-Hinestrosa, R. 1998. INRENARE. Presentation to Yale students on March 20. Altos de Campana National Park, Panama.

Current, D. 1994. Forestry for Sustainable Development: Policy Lessons from Central America and Panama. The Environmental and Natural Resources Policy and Training Project.

Economist Intelligence Unit (EIU). 1995. Country Profile 1994-95: Panama. Economist Intelligence Unit Limited.

Gradwohl, J. and R. Greenberg. 1988. Saving the Tropical Forests. Island Press, Washington, D.C.

Gudeman, S. 1978. The Demise of a Rural Economy: From Subsistence to Capitalism in a Latin American Village. Routledge and Keegan Paul, London.

Hanily, G. 1998. Fundacion Natura. Presentation to Yale students on March 16. Gamboa, Panama.

Heckadon, Moreno S. 1998. Smithsonian Tropical Research Institute. Presentation to Yale students, March 14. STRI, Panama City, Panama.

Heckadon, Moreno S. 1993. The Impact of Development on the Panama Canal Environment. *Journal of Inter-American Studies and World Affairs*. 5(3): 129-149.

Heckadon, Moreno S. and J. Gonzalez. 1985. Agonia de la Naturaleza. Smithsonian Tropical Research Institute, Panama.

Illueca, B.J. 1985. Demografia Histórica y Ecología del Istmo de Panamá, 1500-1945. In S. Heckadon Moreno and J. Espinosa Gonzalez (eds.), Agonía de la Naturaleza: Ensayos Sobre el Costo Ambiental del Dessarollo Panameño, Instituto de Investigacíon Agropecuaria de Panamá and Smithsonian Tropical Research Institute. Panama.

INRENARE. 1997. Mission Statement and Objectives.

Intercarib, S.A. and Nathan Associates, Inc. 1996. Plan de Usos del Sueloy los Recursos Naturales de la Region Interoceanica. Informe II, Documento 1 Final: Plan Regional Para El Dessarollo De La Region Interoceanica.

La Bonga Community Members, 1998. Discussion with Yale students on March 17. La Bonga, Panama.

Lasswell, H.D. 1971. A pre-view of the Policy Sciences. American Elsever, New York.

Laval, E. 1998. Presentation to Yale students on March 15. Agua Buena, Panama.

Linares, O. and A. Ranere. 1980. Adaptive Radiations in Prehistoric Panama. Harvard University, Cambridge, MA.

Leach, M., R. Mearns and I. Scoones. 1997. Environmental Entitlements: A Framework for Understanding the Institutional Dynamics of Environmental Change. *IDS Discussion Paper* 359.

McNeely, J.A., J. Harrison and P. Dingwall (eds.). 1994. Protecting Nature: Regional Reviews of Protected Areas. IUCN–The World Conservation Union.

Mejia, A. 1998. ARI. Presentation to Yale students on March 16. Gamboa, Panama.

Mortimer, L. 1987. Panama: A Country Study. Internet, February 15, 1998. Library of Congress, Washington, D.C. Available http://lcweb2.loc.gov

Pereira, H.C. 1989. Policy and Practice in the Management of Tropical Watersheds. Westview Press, Boulder, CO.

Pinchot G. 1947. Breaking New Ground. Harcourt, Brace and Company. New York.

Posey, D. and W. Balee. 1989. Resource Management in Amazonia: Indigenous and Folk Strategies. Advances in Economic Botany No. 7.

Purseglove, J.W. 1972. Tropical Crops: Monocotyledons. Longman Scientific and Technical, New York.

Reisman, W.M. 1995. Institutions and Practices for Restoring and Maintaining Public Order. *Duke Journal of Comparative and International Law*, 6:1.

Sivaramakrishnan, K. In Press. Modern Forests: Trees and Development Spaces in Southwest Bengal. In Lauara Rival (ed). The Social Life of Trees: From Symbols of Regeneration to Political Artifacts, John Gledhill, Bruce Kapferer, and Barbara Bender, (eds.). Explorations in Anthropology Series. Oxford: Berg.

Wadsworth, F. 1978. Deforestation: Death of the Panama Canal. In Proceedings of the US Strategy Conference on Tropical Deforestation, Washington, DC.

Wells, M. and K. Brandon. 1992. People and Parks: Linking Protected Area Management with Local Communities. The International Bank for Reconstruction and Development.

Zimbalist, A. and J. Weeks. 1991. Panama at the Crossroads: Economic Development and Political Change in the Twentieth Century. University of California Press, Berkeley, CA.

Working Toward Effective Policy Processes in Panama Canal Watershed National Parks

Keely B. Maxwell
Christopher J. Williams

SUMMARY. In the Panama Canal Watershed, the formulation and implementation of national park management policies has yet to realize all of the conservation, recreation, and educational goals set for the parks. We identify two underlying conditions that contribute to this breakdown in the policy process. First, national park management policies are based on the traditional United States national park model rather than on a model tailored to the unique ecological and social context of the Canal Watershed. Second, the structure and dynamics of participating institutions do not support effective decision making processes. This paper utilizes the policy science framework to provide a detailed analysis of these obstacles to effective park management and gives recommendations for improving the objectives for, decision processes concerning, and participant cooperation in national parks in the Canal Watershed. We recommend that participants initiate a prototype management program in one of the national parks in the Watershed that is easily adaptable and can be used as a demonstration site for other national park managers and participants. *[Article copies available for a fee from The Haworth Document Delivery Service: 1-800-342-9678. E-mail address: getinfo@haworthpressinc.com <Website: http://www.haworthpressinc.com>]*

Keely B. Maxwell is a Doctoral Student, Yale School of Forestry and Environmental Studies, New Haven, CT 06511.

Christopher J. Williams recently completed a Master of Forestry degree at the Yale School of Forestry and Environmental Studies, New Haven, CT 06511. He is currently an intern with the Connecticut Nature Conservancy.

[Haworth co-indexing entry note]: "Working Toward Effective Policy Processes in Panama Canal Watershed National Parks." Maxwell, Keely B., and Christopher J. Williams. Co-published simultaneously in *Journal of Sustainable Forestry* (Food Products Press, an imprint of The Haworth Press, Inc.) Vol. 8, No. 3/4, 1999, pp. 147-164; and: *Protecting Watershed Areas: Case of the Panama Canal* (ed: Mark S. Ashton, Jennifer L. O'Hara, and Robert D. Hauff) Food Products Press, an imprint of The Haworth Press, Inc., 1999, pp. 147-164. Single or multiple copies of this article are available for a fee from The Haworth Document Delivery Service [1-800-342-9678, 9:00 a.m. - 5:00 p.m. (EST). E-mail address: getinfo@ haworthpressinc.com].

KEYWORDS. Panama Canal, national park management, policy sciences, natural resource agencies, watershed management, decision making

INTRODUCTION

In the Panama Canal Watershed, as in many other areas of the world, national parks are a critical aspect of natural resource management. According to Panama's Institute of Renewable Natural Resources (INRENARE), the primary motivation for creating these parks in the Watershed has been to preserve the remaining forest cover to ensure high volumes of quality water for Canal operations and human consumption. Other goals include conserving biological diversity, providing opportunities for recreation, scientific research, and education, and preserving scenic and cultural resources (Houseal 1982).

To date, the national parks in the Watershed have been host to many successes. Agroforestry projects with communities in and on the borders in national parks have improved community relationships with INRENARE as well as expedited reforestation of many areas. A partnership has been forged between INRENARE and the Smithsonian Tropical Resource Institute (STRI) to develop a systematic data gathering effort in the Watershed. This Management of Natural Resources (MARENA) project will be useful in understanding the biological and ecological factors that affect the national parks of the Watershed (INRENARE 1997). In addition, INRENARE and other organizations have instituted successful environmental education programs including guided nature trails.

Despite these and other successes, parks in the Canal Watershed are struggling with many difficulties. Problems include conflicts with local community members; internal organizational inefficiency; frustration of natural resource managers; habitat fragmentation within and outside parks that is negatively affecting biodiversity and ecological processes; and continued sedimentation of the Canal. However, these issues are merely symptoms of a larger problem facing Panama's national parks–namely, that current management goals, policies, and strategies, all of which are embedded in institutional dynamics, do not systematically and effectively address obstacles to park success.

To address this larger problem, the participants in national park management in the Watershed need to address two issues. First, participants should define the goals and objectives of parks in the Panama Canal Watershed within the specific context of regional social and biological realities that affect the region. Currently, there is an ill-defined vision of national parks in Panama, a vision largely based on the United States national park model. National parks in the Watershed need to develop a vision or mission based on

Panamanian realities. Second, dysfunctional organizational dynamics and decision processes need to be improved. Natural resource managers and organizations need to develop an effective decision process that incorporates the goals and views of all participants as well as seek ways to integrate their different approaches to natural resource management in the parks.

This paper uses the framework and theories outlined by the policy science approach to problem solving to analyze the issues at hand (Lasswell 1971; Brewer and deLeon 1983; Clark 1992; Clark 1996). The policy science framework provides a logical and methodological approach to solving real-world problems through a thorough analysis of the participants, the decision processes involved in policy making, and the problem itself. However, a detailed analysis of the participants and the problems involved in national park management using the policy science framework would require much more extensive research in the Watershed.[1] Thus, we have confined our analysis to the major participants and their role in decision making in the Watershed to develop recommendations that are also grounded in examples from other parks around the world.

Based on our analysis of Panama's current successes and difficulties and the similarities of their national park problems to other protected area policy dilemmas, we recommend that a prototype park be developed in the Canal Watershed. This prototype would incorporate clearly defined goals and a mission based on Panamanian social and biological realities. It would integrate the views and management approaches of the different participants involved in park management by encouraging a more effective decision process. Ultimately, this prototype park would serve as an educational focal park from which general principles could be learned and later applied to other protected areas in the Watershed.

PARTICIPANTS AFFECTING PARK MANAGEMENT

There are numerous participants interested in protected areas management and conservation in the Canal Watershed. Participants are groups or individuals who have a stake in the outcomes of park management decisions. Participant actions and interactions form part of the context that shapes the formal and informal goals for parks of the Panama Canal Watershed, as well as determines the feasibility of meeting park objectives. To accurately predict outcomes and appropriately modify management policies requires knowledge of the participants involved and their perspectives and goals relative to park issues (Mazur and Clark 1997).

The Institute of Renewable Natural Resources (INRENARE) is the prima-

---

1. This analysis is largely based on two of the largest parks in the Watershed–Soberania and Chagres National Parks.

ry organization responsible for day-to-day park management. This government institute was established in 1986. Its primary goal is to protect natural resources within the context of the country's development needs. INRENARE is divided into four departments responsible for (1) protected areas management and wildlife, (2) forestry, (3) environmental education, and (4) watersheds. In addition, there are regional offices in charge of natural resource activities in different sections of the country. Some INRENARE employees have expressed frustration with the fact that communication among these divisions is not optimal. Factors such as differing departmental objectives and jurisdiction, employee training, and competition for limited resources may be hampering more effective interdepartmental and interregional communication.

INRENARE meets its objectives by a variety of activities such as the creation and implementation of park management plans, patrols against illegal activities, environmental education, agroforestry and reforestation projects, timber concessions, and promotion of national legislation. However, as is common around the world, these activities can be immersed in national politics. Executively appointed INRENARE directors have been asked to fill other positions when there has been disagreement over natural resource policies (Jukofsksy 1991; Wali 1993). Suspicion in the early 1990s that INRENARE employees were supporters of General Noriega lowered their credibility with local communities (Jukofsky 1991).

Communities in and around national parks represent the participants perhaps most enmeshed in the park ecosystem. Although the word *community* implies commonality, in reality the term may refer to a heterogeneous or fragmented assemblage of people (Agrawal and Sivaramakrishnan, in press). Park policies tend to quite significantly, and negatively, affect nearby communities (Ghimire and Pimbert 1997). In Panama, the government has restricted livelihood strategies such as swidden agriculture and cattle ranching, which are thought to have detrimental impacts on protected lands.

A variety of peripheral organizations are involved in national park issues, deliberately or by default, with weighty or inconsequential influence on park management. STRI's mission is to collect and be a repository for primarily theoretical ecological information (Simons 1984), which may have relevance to national park management. STRI appears to have had a negligible role as park policy advisors in the past, but they have provided technical advice in recent projects (INRENARE 1997).

Non-governmental organizations such as the National Association for the Conservation of Nature (ANCON) and The Nature Conservancy have agreements with INRENARE to participate in the management of protected areas. These organizations have been involved in staff training, natural resource monitoring, equipment provision, and park boundary delineation (INRENARE 1997). Fundacion Natura, another nongovernmental organization, distributes funding for selected agroforestry, environmental education, research, and reforestation projects in communities near parks.

The Inter-Oceanic Regional Authority (ARI) is responsible for the reversion of the final Canal Watershed lands controlled by the United States. Its goal is to increase economic opportunities for, improve the world image of, and attract international investment in Panama. Some of these lands being reverted have been designated as potential protected areas while others are slated for development (ARI 1996). Development of reverted lands in the Watershed will place additional pressures on existing protected lands by increasing habitat fragmentation and water consumption.

Other participants include government agencies such as the Ministries of Agriculture and Finance, ecotourism companies, international and local tourists, universities, conservation and aid organizations, private landowners, and shipping companies that utilize the Panama Canal.

DECISION PROCESSES AFFECTING PARK MANAGEMENT

The policy science framework helps to elucidate the multiple steps and pathways involved in national park management decision-making. A decision can be thought of not as a discrete event, but rather as a continuum of processes involving the recognition of a problem; gathering and dispersal of relevant information; promotion of ideas; creation, implementation, and enforcement of plans; evaluation of activities; and termination or alteration of policies (Lasswell 1971). Brewer and deLeon (1983) suggest that all of these activities can be divided into six phases: initiation, estimation, selection, implementation, evaluation, and termination. These phases can occur simultaneously. Moreover, policy making is understood as a process in which each of the above phases interacts dynamically in real world problem solving. Understanding the process through which decisions are made can lead to improvements in conservation policies and practices (Clark 1996). Attaining such an understanding in the Canal Watershed should enhance the ability of park managers to achieve park goals.

Initiation

During the initiation phase of the decision process, a problem is recognized, identified, and placed on the public agenda (Clark 1996). This phase allows for creative thinking about a problem, crude hypothesis testing, and early investigation of concepts and claims surrounding a problem (Clark 1992). The frustration of some INRENARE employees involved in park management refelcts the initiation of problem recognition activities in the Panam Canal Watershed. These employees have become disconcerted by the paucity of integration among departments and organizations, negligible government interest and funding support, and ineffectual park management policies. Some enterprising employees have made individual, small-scale efforts at solving these problems. However, there has been no concentrated endeavor to discover and resolve underlying issues of institutional dynamics and few attempts to place these problems on a public agenda.

Estimation

A problem is defined in more detail in the estimation phase through expert and technical analyses. Scientific investigations are utilized to determine which courses of action are likely to have the greatest impacts or most preferred outcomes. Ideally, during this phase information is gathered, processed, and disseminated to those involved in decision making (Clark 1996). In the Panama Canal Watershed, STRI performs the most intensive information gathering. Some of their research provides information about important changes to the ecology of the Canal Watershed, such as the negative effects forest fragmentation can have on bird populations. However, only one project, researching tree species for reforestation, is tailored toward developing solutions to real-world problems in the Watershed. STRI's research results seem to have had limited impact on national park management. This moderate influence is due in part to the fact that their diffusion of research results is primarily limited to publication in English journals. STRI does provide information and advice to other organizations upon request.

INRENARE also gathers information, but on a much less methodical basis (INRENARE 1995). The Management of Natural Resources (MARENA) project, for which STRI is training INRENARE technicians, is its first systematic data gathering effort. As exemplified by the MARENA project, most information gathering about national parks focuses on biological data. While this knowledge is useful, it fails to capture the complexities of the entire human ecosystem in which the park resides. The term human ecosystem (Machlis et al. 1995) designates the social, as well as biophysical, context of the park, implying that humans and their institutions are an irremovable yet dynamic component of the park ecosystem.

In general, knowledge does not seem to be adequately integrated into other aspects of decision making. For example, government land management in the Canal Watershed is predicated on the prevailing assumption that forest cover is equated with improved Canal waters. However, the actual relationship between trees and water flow is complex and unknown (Windsor 1990).

Selection

In the selection phase of the decision process, policies designed for solving the problem are formulated, debated, and authorized. A choice is made from the various options developed during the estimation phase and a formal or informal policy is selected through enactment of legislation or other means (Clark 1996). Unfortunately, there are problems that impede policy selection in the Panama Canal Watershed at both the national and local level.

At the national level, even though INRENARE is the primary institution responsible for park management, it lacks access to some of the policy and decision making arenas in which policy selection affects national parks. There are numerous policy decisions concerning national parks in which even upper level INRENARE management is not included. For example, the Ministry of Agriculture provides incentives for land conversion to agriculture. Despite the fact that this conversion can occur in park ecosystems, INRENARE has no say as to where in the Watershed these policies are to be implemented.

Park management plans in general are limited in scope. They only have the authority to address issues inside parks, not activities in the broader ecosystem that can affect parks substantially. The state of formal national park policy selection is exemplified by the latest management plan written for Soberania. This plan was written by an agronomist lacking substantial experience in protected area management. According to INRENARE park managers, it offered broad guidelines for potential environmental education opportunities in the park, rather than addressing daily administrative needs.

Local park policies are often tailored by park managers based on management plans and policies designated at the national level. Fortunately, INRENARE park guards and managers are arguably the people most intimately acquainted with activities in and around parks that directly influence park conditions. These managers make informal, on-the-ground management decisions based on their observations. However, there is no explicit mechanism to incorporate their expertise and knowledge into more formal park policies. Moreover, park communities are not usually solicited for opinions or information in policy planning.

Implementation

The implementation phase of the decision process is often the focus of most policy analyses because it is the stage at which selected policies are executed. During this phase, specific programs are developed and applied to the problem at hand (Clark 1992; Clark 1996). In the Canal Watershed, park policies are selectively implemented by park managers. For example, hunting and setting fires are fineable offences. However, collecting the fines is an extremely lengthy process, after which the fines revert back to the national treasury instead of benefiting a particular protected area. Thus, as INRENARE employees have pointed out, this technique is practically ineffective and therefore ignored by managers. While such policy adjustment allows managers to utilize procedures more relevant to the situation at hand, it reflects the fact that there are inconsistencies between broader park goals and policies and actual management issues.

Furthermore, as with parks around the globe, the implementation of park policies is strongly impeded by a lack of available monetary resources (Ugalde 1994). Within the Watershed, park guards until recently had neither uniforms nor necessary field equipment to safely and effectively do their jobs. Zoning was called for by the management plan for Soberania, but realization on the ground has been negligible due in part to unavailable funds.

Evaluation and Termination

The evaluation and termination phases of the decision process are often the most neglected in real world policy formulation. During evaluation, the success of the programs implemented and their effectiveness toward solving the problem at hand are appraised. Once a thorough evaluation has been completed, decisions are made as to whether the programs should be terminated, continued, or altered in order to most effectively address the original problem (Clark 1996). In the Canal Watershed, there are few systematic evaluations of management policies. The appraisals that do occur are for the most part informal, and results are not distributed outside the parks in which the changes were initiated. For example, employees who receive additional training may return to INRENARE with suggestions for improvement. There is, however, no formal mechanism for their evaluative efforts to be heard nor for changes to become incorporated within the institution. Because evaluation is neither systematic nor diffusive, policies are altered or terminated without formal appraisal of their effectiveness at improving park conditions.

SETTING PARK GOALS

Before participants strive to improve the decision process, they should focus on defining the problems with park management. Towards this end, there is one fundamental step toward successful park management in the Panama Canal Watershed that has yet to be mastered. That step is the act of setting goals for national parks that reflect the social, ecological, and historical park context. Goals should be in accord with the common interest, which is the concurrent melding and distillation of disparate participant objectives. This setting of goals provides a base upon which further problem solving must rest (Lasswell 1971).

Establishing goals that are contextual and in the common interest should increase the likelihood that these goals will be attainable for the long term. Currently, multiple participants wield differing visions for national parks in the Panama Canal Watershed. Parks are viewed alternatively as living spaces, potential ecotourism meccas, and water sources. None of these pictures captures all aspects of the park ecosystem and therefore leaves room for conflict and misdirection of policies. Furthermore, many of the terms used in official park goal statements (such as ecotourism, ecosystem or watershed management, sustainable development, participation, and deforestation) have not been explicitly defined or grounded in Panama's particular conservation challenges.

Official park goals (Table 1) are modeled after the traditional United States National Park model. In this model, park environs are not necessarily viewed in a manner compatible with ecological processes (Milne and Waugh 1994). Ecosystem boundaries often are not taken into account in park delineation, as can be seen with the rectangular shape of Soberania National Park. As such, landscape fragmentation outside park boundaries can continue to erode park species populations. Natural changes in the landscape are held at bay, as has resulted from fire suppression policies in Yellowstone National Park (Lichtman 1998).

Another aspect of the United States model is that people are viewed as outsiders to otherwise pristine environments (Milne and Waugh 1994). Human activities are restricted to those deemed compatible with conservation goals, which tend to be of a recreational or research nature (McNeely et al. 1994). Exclusion of specific activities and people from parks necessitates authoritarian enforcement as traditional rights are criminalized. Such techniques have led to resentment and conflicts over rights to natural resources in parks around the world (Wells and Brandon 1992). In Chagres National Park, residents have not been excluded from park territories because park managers decided that complete expulsion would be unfeasible. However, land use rights have been restricted, with prohibitions on hunting and agricultural activities.

Conflicts over natural resources are common when the diverse opinions of local participants are not taken into account (Ghimire and Pimbert 1997). In Thailand, Indonesia, and elsewhere, threats and violence have become the mode of communication between park guards and disgruntled citizens (Wells

TABLE 1. Goals for National Parks in the Canal Watershed

- Preserve a representative sample of tropical rain forest, with its diverse flora and fauna.
- Guarantee the preservation of genetic resources and scenic beauty in the area.
- Offer opportunities for environmental education and interpretation.
- Ensure a future for the water supply in Gatun Lake and Lake Alhajuela.
- Conserve sites of historic value that are part of Panama's cultural patrimony.
- Provide opportunities for recreation and tourism.

---

and Brandon 1992). Panama has not entirely escaped this situation. Regular clashes between guards and hunters occur, and fires have been set in parks. These fires have exacerbated the spread of *paja gringa (Saccharum spontaneum* L.) (see Hammond, this issue), reinforcing the irrevocable ties between social and ecological issues. While it is difficult to know the actual motives of those who set these fires, their actions reflect deeper tensions over the right to control natural resources inside park boundaries. These tensions arise from basic differences in how park resources are valued. Interviews with park community members have revealed that these participants tend to focus on protecting and improving their economic and social status; park managers, on the other hand, are interested in preserving natural resources.

Official goals for Panamanian national parks promote recreation, education, and research as being advantageous to the parks. However, these activities too can take their toll on park resources. For example, visitors trample irreplaceable flora in Zimbabwe (McIvor 1997), and trash litters the Himalayas (Wells and Brandon 1992). The consequences of such purportedly beneficial activities should not be forgotten when calculating the distribution of costs and benefits of park activities.

To attain a more common vision for national parks in the Canal Watershed, one relevant to the context at hand, requires several shifts in perspective regarding park management approaches. For example, some natural resource managers in Panama are moving towards more participatory methods of interaction among participants. Enhancing communication pathways is prerequisite to a thorough examination by all participants of fundamental assumptions being made, implicitly or explicitly, in their definitions of a successful national park.

Other innovative concepts that emphasize the scope, uncertainties, and dynamic nature of park ecosystems and their management should be taken into account when developing individual park policies (Table 2). Further guidance may be found in McNeely et al.'s (1994) discussion of national park management. Organizations involved in park management need to think of

TABLE 2. Park Policy Considerations

- Discern and accommodate ecosystem processes and fluxes.
- Expand participation in policy making and enforcement in an appropriate manner.
- Incorporate concerns and values of more stakeholders into all aspects of the decision process.
- Understand the dynamics and heterogeneity of local communities and adapt relations accordingly.
- Define conservation terminology with respect to local park context.
- Continue to manage for park goals beyond park borders.

people as integral to park systems, rather than outsiders brought in only for enjoyment. Management policies should consider up front the goals and perspectives of all participants, even though all human activities may not be desirable in the end. Attempting to understand and manage for ecological processes, uncertain and complex though they may be, may better promote a healthy human ecosystem.

IMPROVING THE DECISION PROCESS

Once appropriate goals have been established, attention should be diverted to improving the decision making processes of participating institutions. Institutions involved in park issues have not paid enough formal and effective attention to their strengths and weaknesses as organizations and to their ability to meet sustainable goals for natural resource management. The lack of communication and coordination among institutions and the frustration of natural resource managers are testaments to this problem. Focusing on institutional capacities is requisite to full resolution of natural resource conflicts (Thompson and Warburton 1985).

Problem recognition (initiation) should be carried out in a creative, open, reliable, and comprehensive manner. It should be done in a way that works to define and support the common interest, instead of the current practice of only defining problems relative to special interests (Lasswell 1971). Participants should keep in mind that they too comprise part of the ever-changing context of the park ecosystem; thus, they should not act as though they exist in a vacuum (Machlis et al. 1995).

In the Columbia River Basin of the United States, for example, problems traditionally were seen by conservationists as a decline in salmon populations or by power companies as a need for more electricity production. In recent years, participants in this region have realized that the larger issue of the sustainability of the entire ecosystem and necessary tradeoffs among resource uses must be addressed. Towards this end, new means of public involvement and negotiation are being utilized (Lee 1997).

Information gathering and dispersal (estimation) should be reliable, open, and widespread (McDougal and Reisman 1981). In Panama, achieving this goal may mean using existing yet under-exploited resources such as park rangers and community members. Developing new task forces for obtaining information, such as the ongoing MARENA monitoring project, could also play a role. It is also important that the information accumulated emphasize the social as much as the biological.

Improving dispersal of information in Panama would require improved interorganizational communication. Currently, relationships among organizations do not serve to synergize efforts toward national park management.

ANCON, INRENARE, and Fundacion Natura are all involved in agroforestry projects, for example, but there seems to be little sharing of techniques among them (INRENARE 1995).

In the Adirondack Mountains of New York, a more successful strategy for information dispersal exists. Paul Smith's College is serving the role of a forum for facilitating the spread of information about Adirondack issues (Rechlin pers. comm.). Assigning a participant the specific task of obtaining and disseminating all information relevant to the Watershed may close existing knowledge gaps. However, the challenge remains to ensure the relevancy of this information to management issues (Machlis pers. comm.).

Institutional structures should also be better geared to support new learning. It appears that participating organizations currently do not promote internal information flow. To address these issues, organizations should take several steps. One would be to foster mutual trust and respect among employees (Mazur and Clark 1997). The efforts by employees of INRENARE to make personal connections among departments are an excellent start. Another step would be to train people in a manner suitable to their level of education. If people cannot read well, for example, audio-visual training may produce even more professional employees. A third strategy would be to stress interdisciplinary teamwork and skills (Mazur and Clark 1997). Yet another would be to create appropriate and specific pathways along which information would flow.

Policy selection processes should be as inclusive as possible by integrating perspectives and other information from numerous sources (McDougal and Reisman 1981). Policy selection concerning national parks in Panama occurs at both national and local park scales. Currently, national level decision making in Panama relegates protected areas to an inferior position. This subordinate status is apparent in the lack of power and resources allocated to INRENARE. There is also neglect of natural resources as a foundation of the country's economic future, as seen by the development oriented goals for the Watershed lands formerly controlled by the United States. Unfortunately, one aspect of national park management other countries have not inherited from the United States is that park agencies typically have not held the same status relative to other government agencies. In the United States, the National Park Service is a division of the Department of the Interior, which is equivalent in power to a Panamanian ministry.

To be able to effectively manage national parks, INRENARE must have a stronger voice in national governance. The current promotion of INRENARE to authority status represents a positive step towards strengthening its position within the government. Unless INRENARE takes great care in this upward transition, however, other participants worry it may become more top-down and rigid in its approach to management. Lee (1992) postulates that large-scale regulation of the environment can lead to the creation of irrelevant and rigid rules. Smaller scale regulation based on personal knowledge and interaction can be more functional and pertinent. How INRENARE will find this balance of power and flexibility remains to be seen.

At the park level, policy selection procedures should include increased representation of participants, including but not limited to local communities and park rangers. Encouraging forms of community participation in Panama might preclude the existing negative relationships between park management and communities that are corollaries of exclusion from decision making. Some employees of INRENARE are starting to promote the concept of participation, but do not yet fully realize of what participation could entail (see Gardner, this volume).

Increased participation could allow park managers to gain a more comprehensive picture of the forces driving people's land use strategies. In addition, understanding how outside economic pressures and decreasing access to land affect people's actions in national parks would allow park administrators to best manage with, not against local communities. Further exploration of the potential and specifics of participation should be done to avoid common pitfalls such as exacerbation of local power imbalances and strain upon local institutional capacities. There may be local barriers to effective participation in terms of extant community dynamics. It also should be realized that additional skills and resources will be required to address the newly broached issues brought up in participatory discussions.

Implementing policies requires that citizens perceive the body in charge as having authority to regulate as well as the ability to enforce regulations (Lasswell 1971). In Panama, there is a distinct lack of resources to effectively control natural resource exploitation in national parks. There also is a lack of feasible enforcement policies. Exploring local involvement in enforcement may increase effectiveness. In the South Luangwa National Park in Zambia, locals have been employed in wildlife policy enforcement (Wells and Brandon 1992). In Amboseli National Park in Kenya, another project has attempted to preclude enforcement by increasing wildlife extension activities (Wells and Brandon 1992).

The evaluation of park policies and organizations involved in park management should be systematic and comprehensive (McDougal and Reisman 1981). It also should include double loop learning. Mazur and Clark (1997) describe single loop learning as ascertaining and correcting immediate problems. All participants are involved in this basic level of problem solving. For example, ANCON's iguana breeding projects address overexploitation of reptiles in Watershed parks. A higher type of learning is double loop learning, which involves assessing and potentially altering the operating premises of an institution. This more comprehensive type of learning is often constrained by bureaucratic structures that do not promote the modification of institutional policies and practices (Mazur and Clark 1997).

The practice of double loop learning is not widely observed in Panamanian conservation organizations. Fundacion Natura, for example, does not examine precisely how the projects it funds have helped to achieve the goals it holds for biodiversity and sustainable development. Likewise, STRI assumes that basic research will be more successful than applied in achieving conservation ends (Rubinoff 1998), but does not seem to be making all of the necessary connections to make this ideal a reality. Creating internal institutional mechanisms to support change and learning will facilitate this task. Inter- and intra-institutional evaluations do not disperse well, for example, perhaps being perceived as condemnations instead of suggestions for improvement.

Finally, ignoring the termination phase of the policy process can lead to chaos and conflict over uncertainties about how changes will affect participants. Explicitly addressing this phase at the outset can preclude conflicts and foster trust among participants. For example, INRENARE's reforestation project on the borders of Soberania National Park addressed this issue by informing farmers that the agroforestry component of the project would have a limited life span. The community's acknowledgement of that fact was a prerequisite to their participation in the project. Therefore, both sets of participants know what to expect from the other, and future projects should be able to avoid mistrust over past unmet expectations.

There are numerous potential pitfalls that should be avoided because they can impede the successful implementation of any well-meant recommendations. Some examples include the following: (1) It is extremely difficult for organizations to change their management styles and perceptions of the world (Gregg and McGean 1985). For example, promoting the vertical transfer of information in institutions may go against long established grains of communication in Panama. (2) In terms of park management, empowerment of local people may be threatening to some (Pimbert and Pretty 1997). (3) Attempts of organizations in Panama to undertake new mandates may strain resources that are insufficient relative to the task at hand, as when INRENARE attempted to create a Center for Forest Research in 1994 (INRENARE 1995). (4) The national context also must not be neglected (Sanderson 1997). Economic and development pressures at the local and national level, nationwide values, and international relations place constraints as well as provide opportunities for national parks in the Panama Canal Watershed.

DEVELOPING A PROTOTYPE PARK

Because of these potential pitfalls, we recommend that a small group within INRENARE approach the tasks of redefining national park goals and improving decision processes at a local scale. A few employees within INRE-

NARE have already begun thinking about innovative management techniques. With identification of key leaders and moderate organization, they could form a small task force with the goal of implementing such approaches on a small scale. Ideally, they would be allowed to develop a *prototype* model for national park management that would be applied to one of the smaller national parks in the Watershed. In this context, a prototype represents a small scale demonstration of the policy process applied to specific park management issues. Prototyping has been identified as a useful strategy for discovering solutions to problems. The managers of a prototype would be constantly evaluating and adjusting management policies to develop a better program based on their previous experiences (Brunner and Clark 1997). This exercise would provide opportunities for other park managers to observe the successes and failures of the prototype and apply relevant policies and solutions to their particular situations.

The prototype should be developed according to the unique situations affecting the focal park in the Canal Watershed. In other words, in order to appropriately redefine the mission and goals of that park, a thorough understanding of the particular biological and social problems that it faces must be reached. In addition, the successes and failures of other projects in national parks around the world could still be used to generate guidelines on how to approach management problems in Panama. The task force in Panama can learn from other attempts to incorporate ecosystem boundaries and processes in comprehensive park management. For example, several criteria have been used to define the Greater Yellowstone Ecosystem, including large bird and mammal ranges, geological similarities, and natural disturbance regimes (Clark and Minta 1994).

While this prototype can serve as the staging ground for the development of a new vision for park management, it can also be used as an experimental forum for improving the decision process in national park management policy. The recommendations suggested above for improving the different phases of the decision process in Panama can be more easily modified and altered on a much smaller scale through the management of a prototype park than if they were to be implemented at a national scale. Furthermore, the prototype approach functions within the limited financial and human resources available. If the task force is given some autonomy to make innovative changes in one park utilizing alternative management frameworks, the resulting management policies will more likely fit with the unique conditions facing the particular national park.

CONCLUSIONS

National parks in the Panama Canal Watershed currently are unable to fulfill completely their conservation, water production, recreation, and education objectives. External pressures on parks are but part of the problem, which in reality lies in institutional dynamics and an ineffective policy and decision process. The findings of this paper suggest that a more appropriate vision for national parks in the Watershed should be formulated and utilized. Furthermore, institutions must examine and alter their structures and relations to create and support a more organized and effective policy process for successful park management. Ultimately, real-world implementation of our recommendations can best be realized at a small scale and on the local level through the creation of a prototype park that allows for both innovation and experimentation.

REFERENCES

Agrawal, A. and K. Sivaramakrishnan. In Press. Agrarian environments. In: A. Agrawal and K. Sivaramakrishnan (eds.). Agrarian Environment Resources: Resources, Representations, and Rule in India. Duke University Press, Durham, North Carolina.

ARI. 1996. Plan Regional para el Desarrollo de la Región Interoceánica. ARI, Panama.

Brewer, G.D. and P. deLeon. 1983. The Foundations of Policy Analysis. The Dorsey Press, Homewood, Illinois.

Brunner, R.D. and T.W. Clark. 1997. A practice-based approach to ecosystem management. *Conservation Biology* 11:38-58.

Clark, T.W. 1992. Practicing natural resource management with a policy orientation. *Environmental Management* 16(4):423-433.

Clark, T.W. 1996. Policy process and conservation biology. In: S. Beissinger, R. Lacy, A. Lugo, and T. Clark (eds.). Conservation Biology: From Theory to Practice. Oxford University Press, New York.

Clark, T.W. and S.C. Minta. 1994. Greater Yellowstone's Future: Prospects for Ecosystem Science, Management, and Policy. Homestead Publishing, Moose, Wyoming.

Ghimire, K. and M. Pimbert. 1997. An overview of issues and concepts. In: pp. 1-45. K. Ghimire and M. Pimbert (eds.). Social Change and Conservation: Environmental Politics and Impacts of National Parks and Protected Areas. Earthscan Publications Ltd., UK.

Gregg, W.P. and B. A. McGean. 1985. Biosphere Reserves–Their History and Promise. *Orion* 4:41-51.

Houseal, B. 1982. Plan de Manejo y Desarrollo del Parque Nacional de Soberanía. INRENARE, Panama.

INRENARE. 1995. Informe de la república de Panama en relación a los avances logrados para el cumplimiento de la meta al año 2000. INRENARE, Panama.

INRENARE. 1997. The Management of Natural Resources Project. INRENARE, Panama.

Jukofsky, D. 1991. The uncertain fate of Panama's forests. *Journal of Forestry* 89(11):17-19.

Lasswell, H.D. 1971. A Pre-View of the Policy Sciences. American Elsevier, New York.

Lee, K. 1997. Sustainability in the Columbia River Basin. In: pp. 214-238. L. Gunderson, C. Holling and S. Light (eds.). Barriers and Bridges to the Renewal of Ecosystems and Institutions. Columbia University Press, New York.

Lee, R.G. 1992. Ecologically effective social organization as a requirement for sustaining watershed ecosystems. In: pp. 73-90. R.J. Naiman (ed.). Watershed Management–Balancing Sustainability and Environmental Change. Springer-Verlag, New York.

Lichtman, P. 1998. The politics of wildfire: Lessons from Yellowstone. *Journal of Forestry* 96(5):4-9.

Machlis, G. 1998. National Park Service, Washington, D.C. Presentation on culture of scientists and managers in national parks, April 23, Yale School of Forestry and Environmental Studies, New Haven, Connecticut.

Machlis, G.E., J.E. Force and W. Burch. 1995. The human ecosystem as an organizing concept in ecosystem management. University of Idaho, Idaho.

McDougal, M. and W. Reisman. 1981. International Law Essays: A Supplement to International Law in Contemporary Perspective. The Foundation Press, Inc., Mineola, New York.

McIvor, C. 1997. Management of wildlife, tourism, and local communities in Zimbabwe. In: pp. 239-269. K. Ghimire and M. Pimbert (eds.). Social Change and Conservation: Environmental Politics and Impacts of National Parks and Protected Areas. Earthscan Publications, Ltd., UL.

McNeely, J., J. Harrison and J. Dingwall. 1994. Introduction to protected areas in the modern world. In: pp. 1-28. J. McNeely, J. Harrison and P. Dingwall (eds.). Protecting Nature: Regional Reviews of Protected Areas. IUCN, Gland, Switzerland.

Mazur, N. and T.W. Clark. 1997. Are Zoos Sustainable? Clarifying, Analyzing, and Addressing Policy Problems. University of Adelaide, South Australia. Draft.

Milne, R. and J. Waugh. 1994. North America. In: pp. 277-299. J. McNeely, J. Harrison, and P. Dingwall. (eds.). Protecting Nature: Regional Reviews of Protected Areas. IUCN, Gland, Switzerland.

Pimbert, M. and J. Pretty. 1997. Parks, people, and professionals: Putting 'participation' into protected-area management. In: pp. 297-330. K. Ghimire and M. Pimbert (eds.). Social Change and Conservation: Environmental Politics and Impacts of National Parks and Protected Areas, Earthscan Publications, Ltd., UL.

Rechlin, M. 1997. Personal communication. Paul Smith's College. Presentation on the college's role in the Adirondack Mountains, May 4, Paul Smiths, New York.

Rubinoff, I. "Message from the Director." Smithsonian Tropical Research Institute. Online. Internet. 15 August 1998. Available http://www.si.edu/organiza/centers/stri/as01.htm

Sanderson, S. 1997. Ten theses on the promise and problems of creative ecosystem management in developing countries. In: pp. 375-390. L. Gunderson, C. Holling, and S. Light (eds.). Barriers and Bridges to the Renewal of Ecosystems and Institutions. Columbia University Press, New York.

Simons, R. 1984. Ten years later: The Smithsonian international experience since the second world parks conference. In: pp. 712-718. J. McNeely and K. Miller (eds.). National Parks, Conservation, and Development. The Role of Protected Areas in

Sustaining Society, Proceedings of the World Congress on National Parks. Bali, Indonesia, 11-22 October 1982. Smithsonian Institute Press, Washington, DC.

Thompson, M. and M. Warburton. 1985. Knowing where to hit it: A conceptual framework for the sustainable development of the Himalaya. *Mountain Research and Development* 5(3):203-220.

Ugalde, A. 1994. Central America. In: pp. 301-322. J. McNeely, J. Harrison and P. Dingwall (eds.). Protecting Nature: Regional Reviews of Protected Areas. IUCN, Gland, Switzerland.

Wali, A. 1993. Transformation of a frontier: State and regional relations in Panama 1972-1990. *Human Organization* 52(2):115-129.

Wells, M. and K. Brandon. 1992. People and Parks: Linking Protected Area Management with Local Communities. World Bank, World Wildlife Fund, and U.S. Agency for International Development, Washington, DC.

Windsor, D.M. 1990. Climate and Moisture Availability on Barro Colorado Island, Panama: Long-Term Environmental Records from a Lowland Tropical Forest. Smithsonian Institute Press, Washington, DC.

The Prospects
for Integrated Watershed Management
in the Panama Canal Watershed

H. Bradley Kahn

SUMMARY. The Panama Canal is inextricably linked to the Watershed which contains it. Effective policy is critical to manage this resource efficiently. Currently, a myriad of divisions along socio-political, organizational and technical lines are hindering integration of practical knowledge and policy experience. This lack of integration is resulting in inefficient management of the natural resources of the Panama Canal Watershed. This paper will rely on the policy sciences framework to describe, analyze and address this problem. It will begin by highlighting important organizations and their efforts in the Panama Canal Watershed, to understand the social context. Next, the trends and conditions which have shaped the problem will be described, to provide a historical context and help make projections for the future. Finally, alternatives will be presented to address the problem. Some of these options include: prototyping exercises to build practice-based experience, workshops designed to teach integration skills and to provide a common experience base for the participants, and an information coordinator to equalize the flow of information between organizations. *[Article copies available for a fee from The Haworth Document Delivery Service: 1-800-342-9678. E-mail address: getinfo@haworthpressinc.com <Website: http://www.haworthpressinc.com>]*

KEYWORDS. Conservation policy, watershed management, Panama Canal, protected areas

---

H. Bradley Kahn received a Master of Forestry degree from the Yale School of Forestry and Environmental Studies, New Haven, CT 06511. He is currently an intern with the Northern Rockies Conservation Cooperative, Jackson, WY.

[Haworth co-indexing entry note]: "The Prospects for Integrated Watershed Management in the Panama Canal Watershed." Kahn, H. Bradley. Co-published simultaneously in *Journal of Sustainable Forestry* (Food Products Press, an imprint of The Haworth Press, Inc.) Vol. 8, No. 3/4, 1999, pp. 165-180; and: *Protecting Watershed Areas: Case of the Panama Canal* (ed: Mark S. Ashton, Jennifer L. O'Hara, and Robert D. Hauff) Food Products Press, an imprint of The Haworth Press, Inc., 1999, pp. 165-180. Single or multiple copies of this article are available for a fee from The Haworth Document Delivery Service [1-800-342-9678, 9:00 a.m. - 5:00 p.m. (EST). E-mail address: getinfo@haworthpressinc.com].

INTRODUCTION

The Panama Canal is inextricably linked to the Watershed that contains it. Panama depends upon the Watershed's resources for the production of high quality, fresh water to fill reservoirs, for hydropower, drinking water, and to allow for transport of global commerce through the canal. Effective policy for the Inter-Oceanic Region and the Canal Watershed is critical to the efficient management of the resource in the common interest. Within the Watershed, there are a myriad of socio-political, organizational and technical divisions which may impede the integration of practical policy experience. This lack of real or potential integration is resulting in inefficient natural resource management, and exacerbation of additional problems, such as deforestation and sedimentation, related to the long-term sustainability of the Panama Canal Watershed (PCW). One way to address this problem is through the promotion of integrated watershed management. In recognition of this fact, the Panamanian Legislative Assembly created the Panama Canal Authority, which "shall coordinate with the corresponding specialized governmental and non-governmental organizations which have responsibility for, and interests in, the natural resources of the Canal Watershed, and shall approve strategies, policies, programs and projects, both public and private, that may affect the Watershed" (Panama Legislative Assembly, 1997a).

This paper will describe the integration "problem" in the Watershed, analyze it to understand its depth and dimensions, and offer constructive recommendations to address improved integration.

My standpoint is shaped by an undergraduate degree in economics, which often manifests itself in a bias towards high level, market mechanisms as tools to achieve policy goals. Recent academic incursions into ecosystem management and the policy sciences have broadened my perspective about the benefits of participatory programs, particularly in situations where decisions affect a broad range of stakeholders. My interest in natural resource management stems from a respect for natural systems, as well as a belief that humans have a responsibility to balance the development and preservation of these resources, in their common interest.

Definitions

Before the problem is described, it is important to define two terms: "integrated" and "watershed management." "Integrated" is defined as "separate parts united together to form a more complete, harmonious or coordinated entity" (DeVinne 1982). "Watershed management" is defined as "the process of guiding and organizing land and other resource use on a watershed to provide goods and services without affecting adversely soil and water resources" (Brooks et al. 1991). The concept of integrated watershed management combines these two definitions, recognizes the connections between entities within a watershed, and seeks to manage resources within the context of the larger natural and human system.

PROBLEM DESCRIPTION

The Social Process

To understand a problem, it is important to examine the social context in which it exists. Clark (1992) describes, "the largest picture possible of a problem and problem setting is . . . the social process map." Because the world is very complex, understanding the social process that shapes a problem is necessary, if creative solutions are to be found. The social process map includes information about "the participants, their perspectives, the situation, their base values, their strategies of action, and the effects of all these elements" (Lasswell 1970).

Currently, there are more than 30 government agencies with authority in the PCW, and often their missions and perspectives are in conflict (Greenquist 1996). To develop a clearer understanding of the problem related to limited integration, certain aspects of the social process will be highlighted. In particular, three organizations are described: the Panama Canal Authority (PCA), the Inter-Oceanic Regional Authority (ARI), and the Institute of Renewable Natural Resources (INRENARE).

The Panama Canal Authority

The PCA, "is responsible for the management, maintenance, use, and conservation of the water resources of the Canal Watershed" (Panama Legislative Assembly 1997a). In addition, with a stated goal of coordinating organizations in the Watershed, the PCA seems ideally situated to address the problem of integration. However, the organization has only recently been constituted. As a result, many details of its operation are unclear. Since the Panama Canal serves as a major route for world trade, and plays a significant role in the Panamanian economy, there is widespread concern about the transition of the Canal from American to Panamanian hands. To address these concerns, the Chairman of the board of directors for the Panama Canal Commission (PCC), Joe Reeder, assured the world business community that the PCA "will be managed according to sound business practices and [will be] insulated from politics." In addition, one source in Panama described the PCA as "business-oriented," and cited a "profit motive," as driving the organization. When combined with the infancy of the organization, this dominant perspective, motivated by wealth, suggests that the PCA may not be properly situated to promote integration.

The Inter-Oceanic Regional Authority

ARI is the Panamanian agency which is charged with the task of promoting, "the use of the areas and goods reverted, for the creation of wealth, intervening to increase the productive activities of exportation of goods and services, [and] the creation of jobs" (Panama Legislative 1997b). Its primary goal is to replace the substantial economic activity lost with the departure of the Americans. Once the Inter-Oceanic Region has been developed into new productive uses, ARI will disband in 2005. To understand the role ARI plays in the Inter-Oceanic Region, it is important to consider the range of values which appear to motivate the organization.

McDougal et al. (1988) explain that a variety of base values "bear significantly on all future outcomes since they are the means by which the goals or policy aims of every group or individual must be sought." At the moment, ARI appears strongly motivated by two base values, wealth and respect. Just as ARI seeks income for Panama, it is also motivated to increase respect for the country in the world community. Since its primary task involves finding new sources of economic activity, it is logical that wealth would serve as an explicit goal. However, the importance of respect offers an interesting perspective on the situation in the Inter-Oceanic Region. Panamanian labor and economic efficiency has often been unfairly characterized by a poor reputation amongst the world community (Greenquist 1996). With the Canal's transition from American to Panamanian control, this perception, even though it is based on a stereotype, represents a threat to the financial success of the operation. To address this perceived lack of respect, a spokesman from ARI elaborated to my group upon the skills and advantages of the Panamanian workforce and economy. With its focus on planned development, ARI is well-suited for this primary task. However, in terms of the problem of poor coordination, there are two confounding factors: the singularity of its purpose and the limited time of its existence.

The Institute of Renewable Natural Resources

INRENARE is the highest level environmental organization in the Panamanian government. However, its status as an Institute has not awarded the agency the full funding or power granted to Ministries or Authorities. In interviews with land managers, clear references were made to low wage scales, difficulties in moving up the bureaucratic ladder, and lack of incentives to seek advanced training. Because of these challenges, it seems likely that attainment of wealth, power and advanced skill are not the primary motivations behind the actions by the organization or its members. Rather, most of the employees, as well as the agency at-large, seem to be motivated by a general belief that their mission, to protect and conserve renewable natural resources, is a valuable one.

INRENARE's difficulties were exemplified during an informal discussion with a park ranger who conveyed a story about one of his personal experiences. Due to risks from poachers, there is a debate about whether national park rangers should carry guns. The story centered upon an incident where the ranger was confronted by a poacher with a shotgun. Because he was unarmed, the ranger felt "disempowered," and he saw the poacher as a serious threat to his well-being. He believed that by arming the rangers, their power would increase, and they would be more successful at meeting their dual goals of personal safety and resource management. However, past efforts to arm the rangers have not succeeded, as the candidates lacked the proper training necessary to pass the required tests.

As the chief environmental agency, INRENARE may seem like a logical choice to promote integrated watershed management. However, as the previous story illustrated, the organization is in a difficult position in the Panamanian government. With a shortage of resources and power, INRENARE may have a difficult time convincing other, more powerful organizations to share power and coordinate decisions. In addition, it is uncertain whether the required skills and expertise are currently present.

The Decision Process

Problems are not static issues that can be addressed uniformly. Rather, they are dynamic processes that are driven by the context in which they exist. Brewer (1983) describes problems, "as having a 'life,' during which time they emerge, are defined and estimated as to their potentialities, are confronted with strategic statements (policies) and tactical measures (programs) that are meant to reduce or resolve their unwanted consequences, and in time end, stabilize, or worsen as the result of both corrective acts and changes in the problem setting itself." This "life" represents the decision process (Brewer 1983). To develop a clearer understanding, the integration problem in the PCW will be examined in terms of this decision process.

The initial recognition and rough description of a problem represents the initiation phase of the decision process, where the nature and complexity of the issue is generally outlined (Brewer 1983). Amidst the milieu of organizations with authority in the PCW, people are aware that there is an integration problem. This impression is based on responses to the question, "how does integration occur between organizations," which we asked during interviews with resource managers. Managers acknowledged that the lack of integration among organizations resulted in inefficient decisions and weakened protection of natural resources, yet they seemed resigned to accept the situation. This sense of complacency can be counter-productive if it thwarts efforts to develop ideas which help define and resolve the problem. While it is unclear if this is the case in the PCW, it is important to promote creative thinking about the problem, to avoid pitfalls such as simplistic definition or domination of the process by a single organization (Ascher and Healy 1990).

Once the decision process has been initiated, the next step involves estimating the problem through, "systematic investigation . . . and thoughtful assessment of options and alternatives" (Brewer 1983). Within the PCW, it is unclear whether detailed study of the problem has been performed. Ascher and Healy (1990) recommend use of diverse experts, in cooperation with affected participants to help estimate the problem. Based on our questioning of a variety of natural resource managers this expert analysis of the integration problem has not yet occurred. For example, creation of the PCA as a coordinating organization represents one selected alternative to help address the problem. Yet, there are many connotations associated with the word "coordination," and it has not been clearly defined. As a result, this fundamental goal of the organization is uncertain. This is only one example which indicates that more effort needs to be concentrated on estimating the problem.

Even though estimation has not been thoroughly carried out, decisions are being made at all organizational, technical and socio-political levels throughout the Watershed. Currently, all stages of the decision process are occurring simultaneously, often without coordination between the parts. As an American organization operating in Panama, the Smithsonian Tropical Research Institute (STRI) is wary of setting goals based upon Panamanian national policy needs, for fear of causing a political "backlash." As a result, its research agenda is determined by the interests of each scientist, without regard to the social context within which they operate. This individual approach results in efforts which are not as useful to Panama as they otherwise could be. In addition, the majority of the findings are published in English, which render them inaccessible to a large portion of the Panamanian management community.

Another example relates to Fundacion Natura (FN), which funds many different projects throughout the Watershed. Its decision to select and implement an action is based upon the quality of the application, rather than any estimation or significance of a problem. While, FN does evaluate the recipient projects for two years to verify that the original goals are being met, after that there is no further monitoring. As in the previous example, FN represents a resource which is not being used to its fullest potential. By increasing its level of integration into national natural resource decisions, the efficiency of FN's funding efforts could be improved.

In reference to the decision process, goals of the policy process can now be clarified. Currently, a wide variety of government and private organizations are pursuing their own agendas with little integration. The result is that the overall decision process is "fragmented," and disconnected. To address this issue, a first step must include a thorough estimation of the integration problem. To create a "level playing field," terms such as "coordination," "integration," and "watershed management" must be defined practically in the Panamanian context. Once the basic terms are clarified, information can be collected to help define the problem. Some questions to investigate might include: How is the problem affecting the PCW? Who are the participants? Who benefits? Who loses? Which organizations are focusing on different portions of the decision process? How are their actions complementing or contradicting those by other entities? How can these organizations be integrated to increase management efficiency?

Once the integration problem has been better understood, then alternative solutions can be proposed, analyzed and selected. As Panama regains control of the Canal over the next 2 years, it is important to move forward effectively to address the integration problem. In this way, the country can be best positioned to select and implement appropriate actions once the transition occurs. Because the PCW is critical to Panama, it is important to coordinate management to increase efficiency of decision-making. While people are aware of the problem, it has not been adequately defined. Once this has been accomplished, efforts can be made to address it. It is important to note that this is not a simple problem. Trying to coordinate agencies and nongovernmental organizations (NGOs) with different missions is a very challenging task. There is a large body of literature which can be accessed learn from past efforts. One potentially useful case study is the Greater Yellowstone Ecosystem, where efforts at integrated management have been underway for many years, with mixed success (Clark 1998; Primm and Clark 1996).

ANALYSIS OF THE PROBLEM

Clark (1996) explains that "lasting, comprehensive solutions to problems cannot be constructed unless the problems themselves are fully understood and analyzed." In order to develop such an analysis, it is critical to develop a problem orientation, which is based upon knowing the trends in the problem which have preceded the current situation, and the conditions which have shaped these trends. In addition, it is important to consider projections of future problems, based upon the trends and conditions (Clark 1992). Three primary categories, socio-political, organizational, and technical, will be used to facilitate the analysis of critical trends, along with the related conditions and projections. These trends will be used to describe a basic pattern. The goal of this analysis is to illuminate critical patterns by differentiating overall trends from local variations and counter-trends.

Trends in the Problem

Socio-Political Trends

In the past, the Panamanian government has avoided making integrated management decisions at a large scale. Rather, the tendency has been to manage at smaller scales, even if the ultimate authority rests high in the government. There are many examples of this which could be used, but two are provided to clarify the point. During our stay, we were informed that Panama does not have a national postal service. Rather, mail is carried by associated post offices (APO), which are generally organized by towns or private entities. This is not meant to imply that the mail service was inefficient, or even inappropriate, given the realities of the Panamanian system. Rather, it points to one example where a system which is typically managed at a large-scale is broken into many small-scale enterprises.

Another example of a bias against large-scale planning occurred in 1990, when INRENARE failed to convince the Panamanian Legislative Assembly to levy a US$.03 tax on each ton of cargo which is transported through the Panama Canal. This increase in toll, from US$1.76 to US$1.79 for each ton of cargo, would have raised US$9,000,000 for protection of the Canal Watershed (Heckadon 1993). While there are many reasons why the measure failed, it adds to the impression that there is a systematic bias against large-scale management.

Another trend, which has thwarted integration in the PCW, is the United States involvement in the area. As a foreign government operating on Panamanian soil, this presence represents a powerful dividing force. Until the 1960's, when a bridge was built, the US control of the Inter-Oceanic Region virtually divided Panama into two countries. In addition to the physical division, the US presence served as a symbolic divider between the two halves of the country. In many ways, it was illogical to consider ecological boundaries, such as a watershed, given the dominance of this political division. How could the Watershed be truly integrated with the US controlling one part, to the exclusion of Panama, which controls the rest? The US presence in Panama clearly increased the difficulty of managing the Watershed at a large-scale.

Organizational Trends

As with many countries throughout the world, inter-organizational cooperation is thwarted by a multitude of bureaucratic divisions in Panama. One current example of this trend stems from the policies advanced by INRE-NARE and the Ministry of Agriculture (MIDA). As the chief steward of renewable natural resources, INRENARE is actively promoting reforestation throughout the Watershed. Through these efforts, the agency hopes to address problems of sedimentation and decreased water yield, as well as habitat deterioration. At the same time, through the use of financial incentive and other methods MIDA is encouraging clearing forests to develop additional agricultural lands. Both the clearing and reforestation efforts are focusing on the Watershed. While it is important to consider site quality when judging the efficiency of these efforts, it seems likely that if the two agencies are not coordinating, then the promotion of these disparate goals will result in inefficient management.

Technical Trends

One final trend is related to a pattern of shifting land-use management in the PCW. In general, this change was promoted, and in some cases, mandated, without full consideration of the large-scale impacts. A primary example of this occurred recently, when a ban against cutting trees (greater than five years old) was implemented in the PCW (Whelan 1988). While this effort may appear to be a constructive way to address the problem of deforestation, it made few concessions for swidden agriculturalists, who were required to file for permission to clear land. The motivation for the action was to encourage settled agriculture within the Watershed. However, the financial and technical support necessary for this approach to succeed was not included. As a result, people who relied on systems of clearing and fallow were forced to move elsewhere. By mandating a change in land-use practices, without offering alternative ways for people to survive, this effort was acontextual, as it failed to consider large-scale impacts.

Conditions Affecting the Trends

Socio-Political Conditions

It is human nature to trust people over institutions and ideologies. Most people will rely on others with whom they are related, or have had shared experiences, more than anonymous organizations. In Panama, this condition is referred to as "personalismo," which loosely translates into "individuality" (Nyrop 1981). Often, a person's health and safety are viewed as a function of their relationships with other people. This fact can be seen in the relations between godparents and children. These godparents are chosen by the child's biological parents as a way to increase their offspring's connections to other people. In this way, the child's well-being increases (Nyrop 1981). This impression was supported during one interview, when it was explained that decisions, including those about resource management, are often made amongst circles of friends and family, and organizations can be seen as outside of these. This same type of situation can be seen in the Greater Yellowstone Ecosystem (GYE), where local ranchers may resent the intrusion of "outsiders" from the National Fish and Wildlife Service. However, integrated watershed management requires cooperation between people and organizations which otherwise may not have strong ties. This condition partly explains why past attempts at large-scale management, in the PCW as well as the GYE, have not been successful.

The US presence in the Inter-Oceanic Region was shaped by the authority endowed upon it by the Hay-Bunau-Varilla Treaty of 1904. In the Inter-Oceanic Region, then known as the Canal zone, the US was granted, "all the rights, power and authority which the United States would possess and exercise if it were the sovereign to the entire exclusion of Panama" (Nyrop 1981). As previously mentioned, the US control of the Inter-Oceanic Region practically divided Panama into two parts. In the quote above, the phrase, "to the entire exclusion of," indicates the extent to which this control was granted. Given such a condition, the difficulties of implementing integrated watershed management in Panama are clarified.

Organizational Conditions

While it is currently a democracy, Panama has a history of authoritarian government. One lingering effect of this is the concentration of power in a relatively few hands in the government. Several sources suggested to us that the "the President can do anything." While this statement may not be completely true, it does speak to a common perception. One example of this is the construction of a road through the Parque Metropolitano. While development is generally prevented in protected areas, an exception was made in this case. The implied explanation for this was the support for the project from offices high in the government. This centralization of power within the highest levels of the government may leave other, lower level organizations, both inside and outside the government, less secure in their power base.

For example, INRENARE, as an Institute in the Panamanian government, is not granted the same powers as a full Ministry. While there are substantive differences between the two designations, there is also symbolic significance. As the chief agency interested in environmental management, INRENARE is a logical choice to integrate efforts within the Canal Watershed. Yet as an institute, it lacks the political power necessary to guide the decisions of other agencies. Given a difference in missions, INRENARE will be forced into a passive position in the face of opposition from Ministries.

Technical Conditions

From 1950 until 1997, the population in the Watershed increased 650%, from 20,000 to 150,000 people (Heckadon 1993; Mitchell 1997). As people moved into the region, patterns of land-use, traditional land management regimes and customary relations all changed. Until the new arrivals have an opportunity to interact with the other inhabitants, they are more likely to form their own groups based on experiences shared in the places they left. This phenomenon is not exclusive to Panama, as one need only consider "China-town" or "Little Italy" in any major American city to see the same effect. Since people everywhere are more likely to coordinate with others with whom they share common experiences, massive immigration will serve to fragment a region. In terms of integration, this fragmentation can increase the difficulty of implementing programs which focus on ecological boundaries, rather than social ones.

Projections of Future Trends

Projections offer a picture of the future, assuming that current conditions continue to be the critical determinants of trends. The motivation for such an analysis is to understand the impacts current conditions and practices will have on future management, and to consider how this will affect the success-ful achievement of the goal of integrated watershed management.

Integrated watershed management on a large scale is a very difficult task to accomplish, as it requires the coordination of many disparate organiza-tions. As previously noted, human nature favors people over institutions. This condition is not likely to change, and thus, given the current approaches to large-scale management, integration is likely to be a slow, arduous task. One notable exception is the creation of the PCA, to aid in coordination within the watershed. However, since it is a new organization, without clearly defined methods at the moment, it is too soon to predict its impact.

With the US withdrawing from the Inter-Oceanic Region in less than two years, there will be a dramatic change in the Panamanian control of the area. While there are a multitude of organizations established to address this transi-tion, currently there is a great deal of uncertainty about their potential for success. The reversion will leave a vacuum in the power dynamics, and it is critical to learn how this will be filled. To avoid mismanaging the resource, it is important for the organizations to be coordinated. However, as previously mentioned, the effectiveness of the PCA in this task is uncertain. The transi-tion represents an unprecedented opportunity for Panama, but it must work quickly to guarantee the organizational infrastructure exists to manage the change.

Given the current conditions, power is likely to remain concentrated in a few hands in the government. If this remains true, the cooperation required between lower level organizations will continue to be lacking, as each "circles the wagons" around its existing power base.

In addition, population projections point to continued growth in the PCW. From this increase, further fragmentation is likely, as immigrants group themselves based on shared experiences and ancestry. These divisions may serve to hinder future attempts to integrate along organizational boundaries. In Panama, I was often reminded that, "people cannot eat trees." If changes in land-use practices throughout the country do not account for this fact, then the increase in population may have dire environmental consequences, as people seek new methods and locations for subsistence and commercial production.

RECOMMENDATIONS

Alternatives

There are four possible approaches to dealing with the integration problem in the Panama Canal Watershed: the status quo, a top-down approach, a participatory approach, and a mixture of the previous two. Because of the trends and conditions outlined in the previous section, the status quo will be dismissed directly. It is clear that ineffective decisions are being made, and a failure to address the problem will not solve this. Both the top-down and the participatory approaches offer distinct benefits, which will be explored and evaluated before a final strategy is selected.

Top-Down Approaches

Top-down efforts are made at high levels of organization, whether in the national government or in NGOs. In addition, there are approaches which are implemented without seeking input from a wide range of affected stakeholders. There are many different top-down methods to address the integration problem. Some include: the use of an information coordinator, workshops for capacity building, incentives for job training, internships, and regulatory reform.

An information coordinator is a high-level government employee who does not serve within any particular organization. The job of this person involves the collection and dissemination of information amongst all government agencies and interested parties. In this way, information, and with it, power can be equalized within the government.

Workshops for capacity building can be used to "teach knowledge and skills for interdisciplinary problem solving to the staffs of government agencies and NGOs as well as community leaders and interested citizens" (Clark 1998). By targeting the participants, these workshops can be very effective for increasing abilities and approaches used by natural resource managers.

Similarly, incentives for job training are intended to increase the practical skills and knowledge of natural resource managers. One manager explained that additional training was not rewarded in terms of job advancement or salary increase. These disincentives would be addressed by giving people preferential placement, or increased wages in return for increased formal experience. Internships would operate under the same motivation. By increasing the training of managers within the government, job skills will increase and results will also improve.

Regulatory reform refers to high-level mandates which direct how the country operates. One example of this is the National Environmental Policy Act (NEPA) in the United States. This law directs all government organizations to consider the environment when making decisions. A similar law in Panama could theoretically direct agencies to consider a specific set of goals when making decisions.

Participatory Approaches

Participatory techniques rely on collaboration between affected people and institutions to make management decisions. While it is still possible that the ultimate authority rests with one organizations, input is solicited from interested stakeholders. Amongst the myriad of participatory approaches, three will be outlined: Prototyping exercises, workshops for community leaders and demonstration projects. "A prototype is a small-scale, trial intervention in a social or policy system" (Clark 1998). While these exercises are designed to address a problem, their primary goal is to collect information, by learning from experience. By starting at a small-scale, risk can be minimized. At the same time, both successful projects and those which are discontinued can be the source of practical policy learning. It is important to emphasize the learning aspect of the prototype. A project is considered a success if the lessons learned can be conveyed to other programs, to prevent the repetition of similar mistakes.

Workshops for community leaders serve the same function as those for capacity building mentioned previously. They can convey valuable information and skills both to and from community members. By including influential citizens in the workshops, along with higher level government staff, experiences can be shared between people with diverse backgrounds and perspectives. In this way, the workshops can also address the problem of fragmentation between groups.

Demonstration projects, such as the Rio Cabuya project, can offer valuable practice-based knowledge for community members. If these projects are combined with extension, then the information can be passed to a larger audience, throughout the PCW. In this way, beneficial management techniques can be proliferated, while less successful approaches can be altered or abandoned.

Selected Recommendations

A mixture of the recommendations above is required to address the problem related to the lack of integration in the PCW. The goal of these alternatives is to recognize the prevailing conditions, and then either seek to alter them, or work within them as constraints. The second of these two is more likely, given the difficulty of altering people's behavior.

Prototyping offers a distinct opportunity to gain practical experience while moving towards the goal of integrated watershed management. The key is to start small! A small sub-watershed, within the PCW should be chosen. Criteria used for this selection should center upon increasing the likelihood for success of the project. Desirable factors include a diversity of strong community groups and sustainable natural resource management practices. In addition, it is important to try to involve a wide array of governmental and nongovernmental organizations. In this way, these entities can gain practical experience working together. In addition, the people within the organizations will share experiences together, which may create links between previously disparate groups. By starting small, but including as many participants as possible, the risks are minimized, while the potential benefits are maximized. The knowledge gained from the experience can serve as a model for larger-scale programs in the future.

The PCA represents the organization which is best suited to spearhead the integration task. To promote this effort, the board of directors should participate in a workshop aimed at teaching integration methods and knowledge. It is appropriate to begin with the directors in this case, as they are the ones who will be first selected in the organization. The workshops can rely on a case study approach, so that the participants can learn from similar efforts in other areas. Some possible case studies from the United States include the Greater Yellowstone Ecosystem, the Interior Columbia River Basin, the Lake Tahoe Region and the Portland Metro Region. All of these examples offer experience-based learning opportunities about the challenges and benefits of integration. From these workshops, the directors can be well situated to set the proper atmosphere for the other staff, as they are selected.

The use of a coordinator could effectively equalize the flow of information among governmental organizations. To be most effective, this effort could extend outside of the government, to include community organizations as well. In this way, information can flow between entities to prevent inefficient decisions based on a lack of information. This type of effort might be construed as a threat to some people who use information as a source of power. For this reason, the information coordinator would have to be an independent entity, appointed to a long term to minimize the possibility of political biases.

A final strategy would involve the use of workshops to convey information about natural resource policy and integration efforts. Another goal of these workshops would be to bring people from different groups together to share experiences. To be most effective, the participants should span as diverse a set of backgrounds as possible. High level managers could work alongside village leaders. In this way, a common base of experiences could be built, to help break down barriers which might exist between the groups.

CONCLUSIONS

Efficient natural resource management of the PCW is vital to Panama. The country relies on the region for the production of high quality fresh water for both Canal operations and drinking water. Current management of the PCW suffers from the lack of integration between organizations with authority in the area. Integrated watershed management is one alternative to address this problem. It is an approach which recognizes the connections which exist between entities in a watershed, and seeks to manage the resources in a manner which considers these linkages.

Achieving the goal of integration will require a combination of techniques at a variety of levels. These include prototyping, workshops for building capacity and shared experiences, coordination of information flows, and development of human resources. By combining these approaches practically, the underlying conditions may be addressed. In this way, the management of the PCW can move towards its goal of effective, rational policy to result in long-term sustainability in the common interest.

REFERENCES

Ascher, W. and R. Healy. 1990. Natural resource policy making in developing countries. Durham, NC: Duke University Press.

Brewer, G.D. 1983. The policy process as a perspective for understanding. In *Children, families and government*, edited by E. Zigler, S.C. Kagan and E. Klugmand. Cambridge, MA: Cambridge University Press.

Brooks, K.N., P.F. Ffolliot, H.M. Gregersen and J.L. Thames. 1991. Hydrology and the management of watersheds. Ames, IA: Iowa State University Press.

Clark, T.W. 1992. Practicing natural resource management with a policy orientation. *Environmental Management* 4:423-433.

Clark, T.W. 1996. Policy processes and conservation biology. In *Conservation biology: From theory to practice*, edited S. Beissinger, R. Lacy, A. Lugo and T. Clark. New York: Oxford University Press.

Clark, T.W. 1998. "Interdisciplinary problem-solving: Next steps in the Greater Yellowstone Ecosystem." Paper prepared for a conference on the theory and practice of interdisciplinary work, Stockholm, June 1998.

DeVinne, P.B., ed. 1982. The American Heritage Dictionary. Boston: Houghton Mifflin Company.

Greenquist 1996. Panama at a new watershed: Panama Canal maintenance and the environment. *Americas* 4:14.

Heckadon, S. 1993. The impact of development on the Panama Canal environment. *Journal of Inter-American Studies and World Affairs* 3:129-149.

Lasswell, H.D. 1970. The emerging conception of the policy sciences. *Policy Sciences* 1:3-14.

McDougal, M.S., W.M. Reisman, and A.R. Willard. 1988. The world community: A planetary social process. *The UC Davis Law Review* Vol. 21 no. 3.

Mitchell, J. 1997. Deforestation could dry up the Panama Canal. *The Christian Science Monitor* 23 October.

Nyrop, R.F., ed. 1981. Panama: A country study. Washington, DC: US Government Printing Office.

Panama Legislative Assembly. 1997a. Organic law for the Panama Canal Authority. Law No. 19 (June 11).

Panama Legislative Assembly. 1997b. Regional plan for the development of the Inter-Oceanic Region and the general plan for use, conservation and development of the Canal area. Law No. 21 (July 2).

Panama passes bill to regulate future Canal agency. 1997. http://www.pananet.com/pancanal/public/release/latest.htm. (April, 1998).

Primm, S.A. and T.W. Clark. 1996. The Greater Yellowstone policy debate: What is the policy problem? *Policy Sciences* 29:137-166.

Whelan, T. 1988. Will the watershed hold? *Environment* 3:13-40.

Field Trips in Natural Resources Professional Education: The Panama Case and Recommendations

Tim W. Clark

Mark S. Ashton

SUMMARY. Field trips are vital components of professional education. The goal of field trips is to prepare graduates to be broad-based problem-solvers for sustainable management of natural resources in the common interest. Field trips are ideal vehicles to aid development of the kind of professional needed in today's complex and dynamic natural resources environment. For field trips to be successful, students must actively exercise their skills in thinking, observation, management, and technical subjects, in integration of diverse knowledge and experience, and in applying their judgments in applied contexts. We describe a 10 day field trip to the Panama Canal Watershed in March, 1998, to assist policymakers and managers administer natural resources. The 17 students on the field trip came from diverse backgrounds including from

Tim W. Clark is affiliated with the School of Forestry and Environmental Studies, and Institution for Social and Policy Studies, Yale University, New Haven, CT 06511. He is also affiliated with the Northern Rockies Conservation Cooperative, Box 2705, Jackson, WY 83001.

Mark S. Ashton is affiliated with the School of Forestry and Environmental Studies, Yale University, New Haven, CT 06511.

The authors thank the Smithsonian Tropical Research Institute and INRENARE for the very strong in-country support that they received for this course and in particular for the field trip. In particular thanks go to Mr. Taka Hagiwara of JICA (Japanese Aid)/INRENARE and to Ms. Mirei Endara, Director General of INRENARE.

[Haworth co-indexing entry note]: "Field Trips in Natural Resources Professional Education: The Panama Case and Recommendations." Clark, Tim W., and Mark S. Ashton. Co-published simultaneously in *Journal of Sustainable Forestry* (Food Products Press, an imprint of The Haworth Press, Inc.) Vol. 8, No. 3/4, 1999, pp. 181-197; and: *Protecting Watershed Areas: Case of the Panama Canal* (ed: Mark S. Ashton, Jennifer L. O'Hara, and Robert D. Hauff) Food Products Press, an imprint of The Haworth Press, Inc., 1999, pp. 181-197. Single or multiple copies of this article are available for a fee from The Haworth Document Delivery Service [1-800-342-9678, 9:00 a.m. - 5:00 p.m. (EST). E-mail address: getinfo@haworthpressinc. com].

countries in Central America and Indonesia, Peace Corps experience, and other practical experiences. The Canal Watershed supplies all water for canal operations and drinking water for many Panamanians. The Watershed shows all resource conflicts that characterized upland forested watersheds in many countries. Pre-trip preparations are described, as is the field trip itself to numerous sites involving many discussions, to post-trip activities and reports. A basic analytic framework was used to investigate each resource case students studied (e.g., biodiversity conservation, park management, watershed planning). The framework is comprised of a comprehensive set of conceptual categories dealing with people involved in each case, their perspectives, the situation (including biogeographic and ecological features), values, strategies, outcomes, and effects. This framework is described and illustrated in an Appendix. Five recommendations are made to facilitate successful field trips. *[Article copies available for a fee from The Haworth Document Delivery Service: 1-800-342-9678. E-mail address: getinfo@haworthpressinc.com <Website: http://www.haworthpressinc.com>]*

KEYWORDS. Conservation policy, education, field trips, forestry, Panama Canal, protected areas, students

INTRODUCTION

Natural resources education in university professional schools should matriculate students who can contribute to broad-based problem solving for sustainability in the common interest. Students should be knowledgeable and skilled in critical thinking, observation, management, and technical matters. This is challenging because demands of successful practice are dynamic and complex and the status of specialized technical and broad social knowledge and methods are rapidly evolving. Furthermore, people's expectations and social aspirations are everywhere in flux (McDougal et al. 1989). Devising educational experiences to best prepare students for real, practical work-a-day life is challenging in this context. Field trips have always been considered key to the curriculum, and when combined with lectures, problem sets, readings, and workshops they provide vital experience essential for turning students into successful professionals. Field trips put students on the front line in contact with diverse professionals and other people involved in actual management. Management cases are often conflict laden. Field trips offer students the opportunity to learn what is really involved in on-the-ground management, make comparison with other management cases, and hone their own integrative ability, insight, and judgment without the actual real-life costs of being wrong. Field trips therefore serve as important synthetic experiences wherein students apply, discuss, and integrate all previous education into a concentrated experience in the field where real natural resources policy and management takes place.

This paper describes a 10 day field trip to Panama in March, 1998, as an example of the field trip method in professional education, outlines a broad analytic framework useful in field trips to make systematic observations about management cases, compare cases and their contexts, and makes recommendations about the use of field trips as an educational experience.

THE PANAMA FIELD TRIP

Policymakers and managers from INRENARE requested Yale's outside assessment of the issues that they face in forest management and protection of the Canal Watershed. Our field trip was in response to their invitation. Yale University's School of Forestry and Environmental Studies curriculum is a diverse mix of courses, many requiring field trips. The field trip and course described here (i.e., Forest Conservation for Diversity and Productivity) is an advanced, interdisciplinary one focused on improving practical observational, analytic, and integrative skills. The course taught by M. Ashton is a 3 credit course concerned with protecting and maintaining biological diversity of complex forested ecosystems while producing various goods and services. Examples of independent case studies examined in past years concern landscape management of ecoregions in the Pacific Northwest, Venezuela, Belize, and central and southern Mexico. Students are encouraged to take an extended class field trip to these regions. Resource issues in the course usually revolve around some aspect of balancing the sustainable maintenance of biological diversity and ecosystem functioning with improving the social well-being of rural peoples. Each year the class evaluates a particular ecoregion rich in biological diversity but threatened by social conflict and land degradation. In 1998 the course took a field trip to learn about the Panama Canal Watershed and interrelated natural resources management issues. Students were exposed to very diverse topics, sites, speakers, and a new culture. The end product in the course is published papers on both the biophysical and social dimension of a certain place and resources issue.

Field Trip Instructors, Teaching Assistant, and Students–The field trip was lead by M. Ashton, T. Clark, and Jennifer O'Hara. All the instructors have experience in international resources management and policy. M. Ashton is a forest ecologist interested in understanding the underlying biophysical constraints for the evolution and maintenance of diversity in forested ecosystems. He applies his knowledge for the development and testing of silvicultural techniques for restoration of degraded lands, and for the management of natural forests for a variety of services and products. T. Clark is a conservation biologist interested in application of the conservation sciences. He is particularly interested in identifying problem-solving approaches that are most successful in the field to achieve genuine sustainability. Jennifer O'Hara is a doctoral candidate interested in sustainability of natural resources, especially forested ecosystems, and has worked for over six years in Belize.

The 17 students had diverse undergraduate backgrounds and work experiences. Three students were from countries in Central America, one from Indonesia; four American students had Peace Corps experience, and half of them could speak Spanish. Other students had various field experiences. All were master's degree candidates.

Professional education at Yale places primary emphasis on continuing individual growth. The School of Forestry and Environmental Studies expects a student, in the words of former Dean Graves, "to acquire the habit of self-development, with what help we can give. In the final analysis, it is self-education that enables one to continue intellectual growth and lead in thought and practice. What students need is the opportunity to develop themselves in capacity, to acquire knowledge, to think things through, and to form independent judgements" (Yale Bulletin 1998). This fits the educational philosophy behind our field trip and is described by Banner and Cannon (1997) who suggest that educational experiences should be open ended, guided by authoritative instructors, who seek order in learning, promotion of imagination, patience, and character building. There is no simple receipt for successful field trips. Much has been written about professional education and problem-solving in modern society through university and outdoor experiences (Lasswell 1971, Schon 1989, Schon and Rein 1994, Clark and Reading 1994, Sullivan 1995, Ehrlich 1997). And environmental problem-solving has come to dominate many educational programs (Bardwell et al. 1994, Scholz et al. 1997). Experience has shown that field trips are a powerful educational tool (Crompton and Sellar 1981, Phipps 1988, Miles 1991). Field trips, when combined with appropriate courses (Clark et al. in manuscript), case studies (Clark 1986a), workshops (Clark et al. in manuscript), can graduate professionals who are effective on the job (Clark 1986b, 1993).

Panama and the Canal Watershed–The Panama Canal Watershed was the focus of the field trip on management and conservation of forested landscapes. The Canal Watershed supplies all the water for the operation of the canal (the major economic resource of Panama), the drinking water for the cities of Panama and Colon, where 75% of the country's population resides, and water for irrigation of agricultural crops. The Panama Canal Watershed shows a complex mix of resources conflicts that characterize upland forested watersheds in many regions of Latin America, such as land-clearing forced by farmer displacement, loss of soil productivity, and other economic pressures.

The value of the Watershed as a biological reserve is tempered by the fact that major landuse conflicts exist within it. Campesinos or small farmers have cleared large portions of the forest (including the parks) in the Watershed, causing sedimentation in downstream reservoir systems. Economic benefits derived from the Panama Canal have in themselves served to draw people from around the region and promoted increasing urbanization, which, in turn, has spread into the Watershed, causing conflicts over land rights and uses. The land rights of native Central American peoples have yet to be settled in many parts of the Watershed. Since the 1960's three large National Parks have been established in the Watershed. In addition, substantial areas have been zoned by the US military as off limits and to a large extent temporarily removed these areas from agricultural colonization. Now largely protected as parks, their long-term fate appears uncertain. All these issues can seriously compromise the future hydrological and biological integrity of the Canal Watershed. The Canal Watershed is therefore both a convenient and important working field laboratory for the research and education of future resources managers and policymakers as well as an actual proving grounds for innovative management and policy.

Pre-Trip Preparations–The first part of the course is in preparation for the field trip. It is a process of assimilating knowledge and understanding values about the field trip site, in this case the country of Panama, and the Canal Watershed in particular. Students focus on comprehending the relationships between knowledge and values that form the essential context using whatever biological, social, and physical disciplines they command. In this way students come to know the resources, people, and science of a place before visiting actual management cases of that place. This stage provides background for the field trip and ultimately the final integration of all knowledge about how resources issues evolved and how management decisions were made.

To assimilate as much information as is possible before students go on the field trip, the course sequentially progresses through an intensive series of topical biological and social subjects over 8 weeks. The sequence of topics are: historical biogeography; climate, geology, and soils; diversity theory and forest dynamics; the cultural history of the people; population demographics; education, health, and the economy; labor and rural land use; agricultural and forest systems; protected areas development, law, and policy. Each class in the sequence on Panama is compared with Sri Lanka, a region of exceptional biological and cultural diversity, in which M. Ashton has been working for over 15 years. Formal lectures are presented for Sri Lanka, which serve as a guide or model for each student to follow in preparing and presenting a brief on the same topic for Panama. The briefs are compiled as a guide book prior to the field trip, which is then distributed to all participants on the field trip.

The Field Trip-Over the 10 day period, students and instructors visited both sides of the Canal Watershed to observe aspects of all the major resources issues and conflicts. Much of the field trip is in Spanish and translators were essential. The trip itself largely comprised of site visits and interviews with key informants of communities and nongovernment organizations involved in forest protection and restoration including policymakers and managers from INRENARE, the PCC (Panama Canal Commission), and ARI (Authority for Resource Planning); and researchers from the Smithsonian Tropical Research Institute and the University of Panama. At intervals in the evening round-table meetings were carried out including instructors, students, and when appropriate key informants to discuss the resources issues studied previously. During these meetings students consolidate their own more specific ideas for evaluating a particular part of a resources issue, usually emphasizing a bio-physical, social, or policy perspective, or a mix, depending upon the students' academic background, interests, and disciplinary strengths.

USING A CORE ANALYTIC FRAMEWORK ON THE FIELD TRIP

The field trip introduces students and instructors to a new country and culture as well as many natural resources management issues. It is easy to become overwhelmed by these diverse experiences as the class moves from issue to issue and back again over relatively large geographic scales and from speaker and field guide including other diverse hosts, perspectives, and repre-sentatives of many different organizations, some seeking contradictory goals and carrying out different practices. Students and instructors need a "stable frame of reference" to "map" what they are hearing and seeing as they go, and as basis for follow up analysis and reports in class. To best integrate diverse knowledge and experiences on the field trip, students required a broad, but practical analytic framework as described below to "map" the overall context and the management and policy problems studied. This framework permits students to systematically interrogate field experiences and comprehensively organize their experiences as a basis for making recom-mendations to improve management and policy. This approach has been used previously in natural resources management and policy (Clark et al. in manu-script).

Analytic Framework-The natural resources management and policy pro-cess can be broadly conceived as individuals and groups of people in social process carrying out decision making which seeks to address various re-sources problems to find solutions in the common interest. This is true re-gardless of where people live. This process is always focused on two basic questions: (1) How to use natural resources? and (2) Who decides? Answering these questions is often contentious. As a result, using a "stable frame of reference" to analyze, compare, and discuss diverse cases, which vary in detail, is invaluable. Such a framework needs to be organized around social process, decision process, and problems to be broadly useful (Figure 1). Fortunately such a framework already exists and has been used in natural resources management and policy over the last 30 years (Lasswell 1971, Clark 1992, 1997). The framework of integrated categories forces a higher level of disciplined rationality than would otherwise occur. The framework is highly abstract and can be used to map any natural resources policy or management case (Lasswell 1971).

Using this framework moved the Panama field trip from a travelogue or narrow technically focused one to one that was systematic, comprehensive, and contextual wherein students focused their attention on management processes, technical issues, and relevant social and decision process features all at the same time. Having a logically comprehensive stable analytic framework comprised of integrated categories that must be empirically mapped regardless of the specific management details at hand forces students to think, understand, and learn in a genuine interdisciplinary way, that is realistic and practical.

The framework is problem oriented in the sense that the focus is on solving real management and policy problems (Clark et al. 1996a, b). This requires that goals or outcomes desired be specified, such as sustainable management of the Panama Canal Watershed in ways that enjoy long lasting public support. In turn, problems are discrepancies between goals and real or likely states of affairs. Problems must be specified before they can be solved. Common problems include inadequate policy process, insufficient attention to kinds of knowledge needed and quality of the process itself, and the process is insufficiently consensual (Ascher and Healy 1990). Alternatives to address these problems include improved goal clarification, science-based (e.g., more research), practice-based (e.g., better programs), interdisciplinary (e.g., better integration), and other (e.g., prototyping, learning). Alternatives must be evaluated and judged in terms of whether or not they worked or not worked when used in the past in similar occasions? Why, or under what conditions, did it work or not work? And would it work satisfactorily under existing conditions? This problem oriented process should be repeated on an ongoing basis within limits of time and resources.

The framework also focuses on the people interacting in the natural resources management and policy process, as well as their perspectives, the situation of interaction, including geography, ecology, and institutions involved, values involved, strategies they are using to achieve desired outcomes, and long-term effects (see Appendix). There are six outcomes to management and policy and together they make up a complete process. For example, there is an outcome of management and policy implementation.

Both instructors and students need to be aware of their own observational standpoint in examining various natural resources management and policy cases. All human beings no matter how hard they try to be objective and neutral never fully succeed. To the extent that a person is not truly objective and neutral or fully knowledgeable relative to the issue under study, their rationality is diminished. All field trip members were encouraged to continuously clarify their own standpoint relative to the issue under discussion. For example, different field trip members had different standpoints about forest management and the relationship of government to campesinos. Standpoints include student, teacher, advocate, advisor, reporter, decision-maker, scholar, facilitator, concerned citizen, or other. Also field trip members were encouraged to appreciate the tasks they were carrying out when they were in a role, including clarifying goals, determining trends, analyzing conditions, projecting trends, and inventing and evaluating management and policy alternatives. They were also asked to think about the factors that shaped their standpoint, including culture, class, interest, personality, and previous experience.

Applications and Post-Trip Activities-The field trip examined at least 10 interactive social/decision processes (i.e., natural resources management and policy issues) including diverse problems about wildlife conservation, park management, forest management (natural restoration and agroforestry), human drinking water, biodiversity protection, watershed policy, native peoples and campesenos, transfer of the Canal to Panamian authority, and Panamian national environmental strategy and sustainability, and others.

Student papers analyzed several of these processes. The analytic framework previously described and in the Appendix was used explicitly in several of these papers (e.g., Maxwell and Williams paper "Visions and revisions: Working toward effective policy processes in Panama canal watershed national parks").

After returning from the trip, students continued to evaluate their experiences and issues in Panama through round-table discussion, analytic exercises, and presentations in both M. Ashton and T. Clark's courses. Clark teaches courses on "Foundations of Natural Resources Policy and Management" and "Species and Ecosystem Conservation: Developing and Applying a Policy Orientation." Students focus on their own problem oriented case study characterizing the goal of management and policy, and exploring alternatives to recommended. Some students followed the framework closely in their cases whereas other students kept the framework implicit. Students used literature from the library to compare, support, and contrast their work with research and case studies from other places.

Students prepared final papers. Following internal reviews by the faculty of the School of Forestry and Environmental Studies, experts and officials in Panama, and the students themselves, the papers were externally reviewed. Papers that were of high caliber were published together in this volume.

RECOMMENDATIONS FOR FIELD TRIPS

The concentrated experience of a field trip was judged by all involved to be an extremely valuable experience. There are many short-term and long-term educational benefits of field trips. The Panama field trip was longer than most field trips. Our experience on shorter field trips has also been very positive. We believe that field trips are an essential component of professional school education essential to develop the kind of analytic, integrative, and judgmental abilities needed in real-life professional practice. Like agriculture and other resource professions that must integrate biophysical and social factors, forestry and biodiversity conservation, and more broadly environmental management, must be based on an understanding of the human-dominated and natural systems involved. Understanding the context of natural resources management and policy is key to successful professional practice. Possessing a practical conceptual framework to guide research, management, and policy in the field is indispensable for successful professional practice.

Benefits of Field Trips–Based on the Panama experience and other field trips over the last 10 years, we believe that field trips in resource management should focus on issues and roles that resource managers will face in the future. Field trips such as the one organized for Panama are tailor-made for challenging students to use the observational and problem-solving skills learned in the cloistered settings of academia in actual management and policy situations. The analytic framework used in our field trips shows students that successful natural resources management and policy must be flexible and address real problems in a practical adaptive manner. Post-trip analytic and integrative exercises and preparation of a final paper and presentation are essential follow up elements in the total educational experience. Our field trip, unlike most university teaching, uses a general analytic framework that has proven practical in diverse, actual management and policy settings.

The field trip method demonstrates to students that natural resource management is not a linear, reductionistic-like process dominated by rigid valuation and linear programming as taught in some other courses (Erhlich 1997). As a result of these other courses, some students come to us with the impression that real-life problem solving is like solving a mathematical equation. In reality, objective functions are rare if they exist at all. They offer little to resolve fundamental value conflicts. And their assumption of certainty (or manageable uncertainty) in problem solving via prediction and modeling leading to one or a definite set of "right" answers is misleading and of little help in real-life complex management and policy contexts like the Panama Canal Watershed case. We rarely understand fully in advance what we want or value and virtually never are certain about the outcome of our decisions. We feel that the key role of management and policy is to analyze and integrate diverse information to clarify options for decision makers, including explanations of their full costs and benefits in terms of all the values involved, not just economic ones.

Selection of Field Trip Sites–We have devised criteria to aid us in selecting field trip sites. These were developed to maximize benefits of field trips to insure that the field site is a suitable testing ground for a student who is about to become a professional resource practitioner. First, our field trips are designed to survey a range of management and policy cases at one site. We see that the restoration and conservation of forested watersheds and other ecosystems, and the waterways that influence downstream land use, especially in the tropics will become increasingly important as cities expand and their land use demands effect the needs of rural peoples. These lands are going to be increasingly at the center of conflict between urban and rural peoples, elites and peasants, suburbanites, and farmers. Therefore this topic area provides a general theme that can be explored world-wide.

Second, our field trips are designed to put students into complex situations as described above. We feel that the management of these and other landscapes requires professionals who will have little basic information available to them. We find students to be most uncomfortable when faced with a problem which they have little information available to work with, and yet this is the norm when working in many countries and sites. These information poor situations require that professionals possess the ability to assemble, organize, and analyze all existing information, systematically compare and contrast it with other information from more documented situations, and employing practical problem-solving methods to the real management and policy challenges.

Third, our field trips are designed to challenge students to develop practical and specific problem-solving skills, and social skills to deal with multiple stake-holders in real-life contexts. Future resource managers must learn to accommodate, adapt, and learn to meet the ever-changing values and demands of pluralistic society.

Fourth, our field trips are designed to elucidate the students that "recipe-like" approaches to managing resources will not be workable or acceptable because of many unique site-specific contextual variables. This is the reason we use the broad analytic framework described in Appendix in the field to understand diverse management and policy problems and to guide invention of recommendations to successfully address these problems.

Lastly, our field trips are designed to aid students in obtaining an understanding of how problem-solving can be done with limited and varied capital. This is an important component of understanding the practical limitations to resource management. Foresters and others involved will have to work with a variety of capital sources to raise investment needed for projects. Much of this investment will have to be done by local people, their local governments, and increasingly by private organizations, some for-profit and some that are not-for-profit.

REFERENCES

Ascher, W., and R. Healy. 1990. Natural resources policymaking in developing countries. Duke University Press, Durham, North Carolina.

Banner, J.M., Jr., and H.C. Cannon. 1997. The elements of teaching. Yale University Press, New Haven, Connecticut.

Bardwell, L.V., M.C. Monroe, and M.T. Tudor, eds. 1994. Environmental problem solving: Theory, practice, and possibilities in environmental education. North American Association for Environmental Education, Troy, Ohio.

Clark, T.W. 1986a. Case studies in wildlife policy education. Renewable Resources Journal 4(4):11-16.

Clark, T.W. 1986b. Professional excellence in wildlife and natural resource organizations. Renewable Resources Journal 4(2):8-13.

Clark, T.W. 1992. Practicing natural resource management with a policy orientation. Environmental Management 16:423-433.

Clark, T.W. 1993. Creating and using knowledge for species and ecosystem conservation: Science, organizations, and policy. Perspectives in Biology and Medicine 36:497-525.

Clark, T.W. 1997. Conservation biologists in the policy process: Learning how to be practical and effective. Pp. 575-597 (Chapter 17) in G.K. Meffe and C.R. Carroll, eds., Principles of Conservation biology, 2nd ed. Sinauer Associates, Sunderland, MA.

Clark, T.W. Unpublished manuscript a. Conservation biology and leadership in public policy: Professinal education in the public interest. Conservation Biology (in review).

Clark, T.W. Unpublished manuscript b. Interdisciplinary problem solving in natural resources management and policy: Enhancing professional skills. Journal of Wildlife Management (in review).

Clark, T.W., and R.P. Reading 1994. A professional perspective: Improving problem solving, communication, and effectiveness. Pp. 351-369 in T.W. Clark, R.P. Reading, and A.L. Clarke, eds., Endangered species recovery: Finding the lessons, improving the process. Island Press, Washington.

Clark, T.W., A.P. Curlee, and R.P. Reading. 1996a. Crafting effective solutions to the large carnivore conservation problem. Conservation Biology 10:940-948.

Clark, T.W., D. Glick, and J. Varley. 1996b. Balancing scientific, social, and regulatory concerns in biodiversity management. Pp. 630-646 in R.C. Szaro and D.

Johnston, eds., Biodiversity in managed landscapes: Theory and practice. Oxford University Press, New York.

Clark, T.W., and R.L. Wallace. 1997. Understanding the human factor in endangered species recovery: Introduction to human social process. Endangered Species Update 15(1):2-9.

Clark, T.W., R.J. Begg, and K. Lowe. In Press. Interdisciplinary problem-solving workshops for Flora and Fauna Branch Professionals, Department of Conservation and Natural Resources, Victoria, Australia. Yale School of Forestry and Environmental Studies, Bulletin Series.

Clark, T.W., A.R. Willard, and C.R. Cromely, eds. Unpublished manuscript. Foundations of natural resources policy and management. Book manuscript submitted to Yale Press, New Haven, Connecticut.

Crompton, J.L., and C. Sellar. 1981. Do outdoor education experiences contribute to positive development in the affective domain? Journal of Environmental Education 12(4):21-29.

Ehrlich, P.R. 1997. Ecologists, rewards, and public education. Pp. 167-173 in A world of wounds: Ecologists and the human dilemma. Excellence in ecology No. 8, O. Kinne, editor, Ecology Institute, D-21385 Oldendorf/Luhe, Germany.

Lasswell, H.D. 1971. Professional training. Pp. 132-159 in A pre-view of policy sciences. American Elsevier, New York.

McDougal, M.S., W.M. Reisman, and A.R. Willard. 1989. The world community: A planetary social process. University of California, Davis Law Review 21(3):807-972.

Miles, J.C. 1991. Teaching in wilderness. Journal of Environmental Education 22(4):5-9.

Phipps, M. 1988. The instructor and experiential education in the outdoors. Journal of Environmental Education 20(1):8-16.

Scholz, R.W., B. Fluckiger, R.C. Schwarzenbach, M. Stauffacher, H. Mieg, and M. Neuenschwander. 1997. Environmental problem-solving ability: Profiles in application documents of research assistants. Journal of Environmental Education 28(4):37-44.

Schon, D.A. 1989. Implications for improving professional education. Pp. 303-326 in Educating the reflective practitioner. Jossey-Bass Publishers, San Francisco, California.

Schon, D.A., and M. Rein. 1994. Conclusion: Implications for research and education. Pp. 188-209 in Frame reflection: Toward the resolution of intractable policy controversies. Basic Books, New York.

Sullivan, W.M. 1995. What is professional knowledge? Expertise and professional education. Pp. 159-190 in Work and integrity: The crises and promise of professionalism in American. HarperBusiness, New York.

Willard, A.R., and C. Norchi. 1993. The decision seminar as an instrument of power and enlightenment. Political Psychology 14:575-606.

Yale School of Forestry and Environmental Studies. 1998. School of Forestry and Environmental Studies 1998-1999. Bulletin of Yale University Series 94 Number 2.

APPENDIX

A framework for interdisciplinary questions to ask in social and decision
process mapping of a specific natural resources problem
(Lasswell 1971, Willard and Norchi 1993, Clark and Wallace 1997,
Clark in manuscript a,b).

---

1. Participants

Who is participating?

Identify both individuals and groups.

Whom would you like to see participate?

Who is demanding to participate?

2. Perspectives

What are the perspectives of those who are participating?

Of those you would like to see participate?

Of those making demands to participate?

What would you like their perspectives to be?

Perspectives include:

A. **Demands** or what participants or potential participants want, in
terms of values and organization.

B. **Expectations** or the matter-of-fact assumptions of participants
about past and future.

C. **Identifications** or on whose behalf are demands made?

3. Situations

In what situation do participants interact?

In what situations would you like to see them participate?

4. Base Values

What assets or resources do participants use in their efforts to achieve
their goals?

All values, including authority, can be used as bases of power; these
values are:

A. **Power** is to make and carry out decisions.

B. **Enlightenment** is to have knowledge.

C. **Wealth** is to have money or its equivalent.

D. **Well-being** is to have health, physical and psychic.

E. **Skill** is to have special abilities.

F. **Affection** is to have family, friends, and warm community relationships.

G. **Respect** is to show and receive deference.

H. **Rectitude** is to have ethical standards.

What assets or resources would you like to see participants use to achieve their goals?

5. **Strategies**

What strategies do participants employ in their efforts to achieve their goals?

Strategies can be considered in terms of diplomatic, ideological, economic, and military instruments.

What strategies would you like to see used by participants in pursuit of their goals?

6. **Outcomes.** There are six kinds of outcomes, one for each of the decision process phases: initiation, estimation, selection, implementation, evaluation, and termination. Outcomes also refer to the ways in which values are shaped and shared. Ask:

What outcomes are achieved in the ongoing, continuous flow of interaction among participants overall and by phase?

Outcomes can be considered in terms of changes in the distribution of values.

Who is indulged in terms of which values?

Who is deprived in terms of which values?

The particular ways in which values are shaped and shared are called practices or institutions.

How are practices changing?

How would you like to see practices change?

What is your preferred distribution of values?

(1) **Initiation:** a problem is perceived, identified, and placed on the public agenda. Creative thinking and intelligence gathering are necessary; simple hypothesis testing takes place; and preliminary look at concepts and claims is undertaken. Outcomes are the putting of information about the problem on the public agenda, including possible initial problem definitions and proposals. Standards include: reliability, comprehensiveness but selective, creative, and open. Ask:

> How did the issue originate?
>
> Who first framed it for other participants to address?
>
> Was the issue identified in a timely way?
>
> Who's interests are favored by an initial problem definition?
>
> How would you like to see initiation proceed?

(2) **Estimation:** the problem is defined in more detail using expert analysis and technical considerations. Open debate is required. Scientific investigations set plausible options for responding based on likely impacts and outcomes. Outlines of a programmatic response are looked at. Performance indicators are crudely listed. Critical parameters are listed. Outcomes are the gathering, processing, and dissemination of information for decision, including alternative policies. Standards are: rational, integrated, comprehensive, and effective. Ask:

> Is information being collected on all relevant components of the problem and its context and from all affected people?
>
> To whom is information being communicated?
>
> How is it used?
>
> Which groups (official or unofficial) urge which courses of action?
>
> What values are promoted or dismissed by each alternative and what groups are served by each?
>
> How would you like to see the estimation phase carried out?

(3) **Selection:** a policy response to the problem is formulated, debated, and authorized by a legitimate source. A prescription is produced. Debate is focused on the actual issues. A choice is made about options and uncertainty is reduced. Outcomes are the formal or informal policy which stabilize expectations on the rules to be enforced under various circumstances, including but not limited to enactment of legislation. Standards are: comprehensive, rational, and open. Ask:

Will the new prescriptions harmonize with rules by which the participants and institutions already operate, or will they conflict?

What rules does the group set for itself and others?

What parts of the prescription are binding and which are not (these are easier to determine if they are written down)?

How would you like to see the selecting phase carried out?

(4) **Implementation:** a program is developed and applied to the problem. The programs relationship to existing institutions is defined. Costs are minimized. Performance expectations are detailed. Both enforcement and dispute resolution are required. Outcomes are the final characterization of a specific instance or case relative to the prescription, including policing and other means. Standards are: timely (prompt), open, dependable in characterizing facts, rational, not open to abuse by individual members, effective, rational (conforming to common interest prescriptions), uniform (independent of special interests), effective (must work in practice), and constructive (mobilizing consensus and cooperation). Ask:

Is implementation consistent with prescription?

Who should be held accountable to follow the rules?

Who will enforce the rules prescribed in selection?

How would you like to see enforcing activity carried out?

Will disputes be resolved by people with authority and control?

How do participants interact and affect one another as they resolve disputes?

How would you like to see implementation carried out?

(5) **Evaluation:** ex post appraisal of the implementation effort and the original policy formulation. A comparison is made between the estimated performance levels with those actually obtained in implementation. Outcomes are the appraisal of the flow of decisions relative to the prescriptions (goals), and identification of those parties formally and informally responsible for successes and failures. Standards are: dependable, realistic, ongoing, independent of special interests, fully contextual (taking many factors into account, including matters of rationality, politics, and morality). Ask:

Who is served by the program and who is not?

Is the program evaluated fully and regularly?

Who is responsible and accountable for success or failure?

By whom are one's own activities appraised?

How would you like to see evaluation carried out?

(6) **Termination:** discontinuation, revision or success of policy. Outcomes are the ending of a prescription and a focus on claims of people who acted in good faith when the prescription was in effect. Standards are: prompt, respectful and consistent with human dignity principles, comprehensive, balanced, and ameliorative. Ask:

Who should stop or change the rules?

Who is served and who is harmed by ending a program?

How would you like to see termination carried out?

7. **Effects.** Ask:

What the long term effects on the social and decision process involved?

What new practices have been put into place?

Were there any innovations?

How were these diffused or restricted?

Panama Canal Watershed:
A Synthesis

Mark S. Ashton
Jennifer L. O'Hara

At the close of this century the future of the Panama Canal will be officially in the hands of the country of Panama. This collection of papers represents an apt review of the status of the Canal at a time when Panama is undergoing a major strategic planning exercise concerning the lands of the Canal Watershed. Our evaluation and commentary is meant to provide a perspective that freshens the debate about the issues in land use concerning the Watershed's sustainabilty in supplying the various services so important to the survival of the country. Our papers are in no way supposed to be comprehensive reviews of different issues within the Watershed. There are many aspects of the Watershed that have been ignored such as evaluating the ecological framework to preserve biodiversity, or investigating the validity of developing commercial markets for tree plantations of teak and mahogany.

Our papers are, however, intended to inject new ideas and evaluations that cover certain biophysical phenomena and in particular the organizational structures and policies that govern the management of the Canal Watershed. On the biophysical side authors have clarified some of the issues on the roles of invasive exotics within the Watershed, in particular *Saccharum spontaneum* (elephant grass), and suggest alternative silvicultural treatments that catalyze forest restoration based on studies from elsewhere in the tropics.

Mark S. Ashton is Associate Professor of Silviculture, School of Forestry and Environmental Studies, Yale University, New Haven, CT 06511.

Jennifer L. O'Hara is Doctoral Candidate, School of Forestry and Environmental Studies, Yale University, New Haven, CT 06511.

[Haworth co-indexing entry note]: "Panama Canal Watershed: A Synthesis." Ashton, Mark S. and Jennifer L. O'Hara. Co-published simultaneously in *Journal of Sustainable Forestry* (Food Products Press, an imprint of The Haworth Press, Inc.) Vol. 8, No. 3/4, 1999, pp. 199-201; and: *Protecting Watershed Areas: Case of the Panama Canal* (ed: Mark S. Ashton, Jennifer L. O'Hara, and Robert D. Hauff) Food Products Press, an imprint of The Haworth Press, Inc., 1999, pp. 199-201. Single or multiple copies of this article are available for a fee from The Haworth Document Delivery Service [1-800-342-9678, 9:00 a.m. - 5:00 p.m. (EST). E-mail address: getinfo@haworthpressinc.com].

Much of these grasslands could be reforested using agroforestry as a medium for initiating partnerships between local villages and public agencies. Several papers suggest criteria that center on co-management as an avenue to be explored. The criteria discussed may facilitate government objectives of re-forestation for protecting the Watershed with local values that seek employment and income from the land.

Larger scale considerations in forest protection that have been suggested are directly related to investments that should be made in the services that the Canal Watershed provides. Carbon offset investments, sedimentation control programs, and ecotourism are three such areas that we have reviewed. All have the capacity to serve as significant reasons for protecting the Watershed. Currently, there is uncertainty about a global carbon tax or quota system that might induce carbon trades from which the protection of forest in the Panama Canal Watershed could benefit. But future foreign investment for carbon sequestration could well develop into an attractive investment for power companies and should be pursued.

Sedimentation problems and water quality issues are, however, much more pressing and are now related to the ability of downstream users to directly comprehend and take action to protect their own Watershed. In the last several years, cities in the northeastern United States that are dependent on surficial water supplies, have actively invested billions of dollars to avoid the costs of building filtration plants. For example, water for the city of New York no longer meets EPA standards. The city was faced with building a filtration plant at a cost of $6-$8 billion along with an additional annual cost of $300,000 million for maintaining and operating the plant (Chichilinisky & Heal 1998). New York wisely decided to restore the watershed of the Cats-kills which is expected to be about a $1-$1.5 billion investment–and these calculations are conservative because this is a consideration made only for drinking water. Other cities have made similar financial decisions, including those with filtration plants, such as Boston, Philadelphia, and Washington, DC. The citizens of Boston and the Massachusetts Water Resources Author-ity are now under a $200 million plus mandate to build a filtration plant from the EPA if they do not continue to improve the quality of drinking water (Howe, 1998).

Though myths abound about the possibility of dramatically increasing water supply through reforestation of the Canal Watershed–there are no such myths about the protection of existing forests and the potential for reforesta-tion of treeless areas to improve water quality and to better control water flows for downstream users. This is becoming very clear. Panama can learn from the problems that other cities are now facing with issues relating to drinking water alone being an economic factor that can dictate the cost-bene-fit analysis of preserving the Watershed from development. Apart from the drinking water issue, the other benefits for preserving the Canal Watershed for extending the operable life of the canal, carbon sequestration, long-term sustainability of landuse generated by better practices in agriculture and forestry, and ecotourism only add to this argument.

Over half of the book evaluates the organizational structure and behavior of different institutions and stakeholders that manage the Watershed. A mixture of recommendations have been made that address the age-old problems of integration among organizations that were originally mandated to manage different resources that often conflict with each other. Authors of these papers ask such questions as what is integrative watershed management? Not that a well defined answer exists in reality. But suggestions have been made for: (i) a participatory planning process that incorporates local users within the context of mandates from government organizations; (ii) prototyping new management strategies on a single site before embarking on the whole Watershed; and (iii) developing an organizational infrastructure for sharing information easily.

In summary, we make a strong case for investing in the Canal Watershed area. Maintaining the integrity of the system in the long run will provide the greatest good for the greatest number of people. An economic evaluation of the forest and hydrological services provided by the Watershed, if done carefully, will we think be a surprise to those critics who consider the development of the Canal Watershed as Panama's only alternative to future prosperity.

REFERENCES

Chichilinisky, G. & G. Heal. 1998. Economic returns from the biosphere. Nature 391: 629-630.
Howe, P.J. 1998. EPA leaders plead for OK on filtration: Pressure's on MWRA in water treatment vote. Boston Globe Newspaper, October 21st.

Index

Agriculture. *See* Agroforestry;
  Saccharum spontaneum
  (Gramineae); Small farmer
  migration
Agroforestry
  small farmer migration and,
    11-12,18-20
  See also Agroforestry case study
    assessment; Carbon
    sequestration forest
    conservation project; Public
    participation in natural
    resource management;
    Saccharum spontaneum
    (Gramineae)
Agroforestry case study assessment
  analysis and recommendations
    regarding, 48-50
  case studies and, description of
    Agua Buena, 43-45,48
    La Bonga, 46-48
    Rio Cabuya, 45-46,48-49
  conclusions regarding, 50
  introduction regarding
    agroforestry, importance of, 40
    definitions used in, 40-41
    evaluation criteria and, 41-43
  small farmer participation rates
    and, 41-42
    community participation factors
      and, 43
    economic feasibility factors and,
      42-43
    extension services factors and, 43
    incentives to promote, 42
    technology factors and, 42
  summary regarding, 39
Agua Buena Agroforestry project, 6
  community participation in, 45

economic feasibility of, 44-45
  incentives behind, 44
  management objectives of, 44
  project life span of, 44
  summary analysis of, 48
  technology used at, 44
Alhajuela Watershed, 2,3*fig.*,4
  description of, 83-84,83*fig.*
  sediment yield of, 81,84,86
  vs. US reservoirs and,
    86-89,87*fig.*,88table
  water demand from, 82
Altos de Campana National Park, 4,55
Alvarado, Luis, 6
Amboseli National Park, Kenya, 159
ANAM (National Environmental
  Authority), xii
ANCON. *See* National Association for
  the Conservation of Nature
  (ANCON)
ARI. *See* Inter-Oceanic Region
  Authority (ARI)
Arias, Ricardo Alberto, 73
Ashton, Mark, 183,188,199

Balboa West Range, 71,72*fig.*,75
Barletta, Nicolas Ardito, 95
Barro Colorado Island, 121
Barro Colorado National Monument,
  108
Block, Nadine, 53
Boqueron River sub-watershed,
  86,87*fig.*,88*table*

Cabuya de Chilibre reforestation
  project, 34
Cabuya de Chilibre village, 6